The **BEST** **WRITING** on **MATHEMATICS**

2013

The BEST WRITING on MATHEMATICS

2013

Mircea Pitici, Editor

FOREWORD BY
ROGER PENROSE

PRINCETON UNIVERSITY PRESS
PRINCETON AND OXFORD

Copyright © 2014 by Princeton University Press
Published by Princeton University Press, 41 William Street,
Princeton, New Jersey 08540
In the United Kingdom: Princeton University Press,
6 Oxford Street, Woodstock, Oxfordshire OX20 1TW

press.princeton.edu

All Rights Reserved

ISBN (pbk.) 978-0-691-16041-2

This book has been composed in Perpetua

Printed on acid-free paper. ∞

Printed in the United States of America

1 3 5 7 9 10 8 6 4 2

For William P. Thurston
In memoriam

Contents

Foreword

Roger Penrose

Although I did not expect to become a mathematician when I was growing up—my first desire had been to be a train driver, and later it was (secretly) to be a brain surgeon—mathematics had intrigued and excited me from a young age. My father was not a professional mathematician, but he used mathematics in original ways in his statistical work in human genetics. He clearly enjoyed mathematics for its own sake, and he would often engage me and my two brothers with mathematical puzzles and with aspects of the physical and biological world that had a mathematical character. To me, it had become clear that mathematics was something to be enjoyed. It was evidently also something that played an important part in the workings of the world, and one could see this basic role not only in physics and astronomy, but also in many aspects of biology.

I learned much of the beauties of geometry from him, and together we constructed from cardboard not only the five Platonic solids but also many of their Archimedean and rhombic cousins. This activity arose from one occasion when, at some quite early age, I had been studying a floor or table surface, tiled with a repeating pattern of ceramic regular hexagons. I had wondered, somewhat doubtfully, whether they might, if continued far enough into the distance, be able to cover an entire spherical surface. My father assured me that they could not but told me that regular pentagons, on the other hand, would do so. Perhaps there was some seed of a thought of a possible converse to this, planted early in my mind, about a possibility of using regular pentagons in a tiling of the plane, that found itself realized about one third of a century later!

My earliest encounter with algebra came about also at an early age, when, having long been intrigued by the identity $2 + 2 = 2 \times 2$, I had hit upon $1\frac{1}{2} + 3 = 1\frac{1}{2} \times 3$. Wondering whether there might be other examples, and using some geometrical consideration concerning squares and rectangles, or something—I had never done any algebra—I hit upon some rather too-elaborate formula for what I had guessed might be a general expression for the solution to this problem. Upon my showing this to my older brother Oliver, he immediately showed me how my formula could be reduced to $\frac{1}{a} + \frac{1}{b} = 1$, and he explained to me how this formula indeed provided the general solution to $a + b = a \times b$. I was amazed by this power of simple algebra to transform and simplify expressions, and this basic demonstration opened my eyes to the wonders of the world of algebra.

Much later, when I was about 15, I told my father that my mathematics teacher had informed us that we would be starting calculus on the following day. Upon hearing this, a desperate expression came over his face, and he immediately took me aside to explain calculus to me, which he did very well. I could see that he had almost deprived himself of the opportunity to be the first to introduce to me the joy and the magic of calculus. I think that almost as great as my immediate fascination with this wonderful subject was my father's passionate need to relate to me in this mathematically important way. This method (and through other intellectual pursuits such as biology, art, music, puzzles, and games) seems to have been his only emotional route to his sons. To try to communicate with him on personal matters was a virtual impossibility. It was with this background that I had grown up to be comfortable with mathematics and to regard it as a friend and as a recreation, and not something to be frightened or deterred by.

Yet there was an irony in store for me. Both my parents had been medically trained and had decided that of their three sons, I was the one to take over the family concerns with medicine (and, after all, I had my secret ambition to be a brain surgeon). This possibility went by the wayside, however, because of a decision that I had had to make at school that entailed my giving up biology in favor of mathematics, much to the displeasure of my parents. (It was my much younger sister Shirley who eventually took up the banner of medicine, eventually becoming a professor of cancer genetics.) My father was even less keen when I later expressed the desire to study mathematics, pure and simple, at university.

He seemed to be of the opinion that to do *just* mathematics, without necessarily applying it to some other scientific area of study, one had to be a strange, introverted sort of person, with no other interests but mathematics itself. In his desires and ambitions for his sons, my father was, indeed, an emotionally complicated individual!

In fact, I think that initially mathematics alone was my true main interest, with no necessity for it to relate to any other science or to any aspect of the external physical world. Nor had I any great desire to communicate my mathematical understandings to others. Yet, as things developed, I began to feel a greater and greater need to relate my mathematical interests to the workings of the outside world and also, eventually, to communicate what understandings I had acquired to the general public. I have, indeed, come to recognize the importance of trying to convey to others an appreciation of not only the unique value of mathematics but also its remarkable aesthetic qualities. Few other people have had the kind of advantages that I have myself had, with regard to mathematics, arising from my own curiously distinctive mathematical background.

This volume serves such a purpose, providing accounts of many of the achievements of mathematics. It is the fourth of a series of compendia of previously published articles, aimed at introducing to the general public various areas of mathematics and its multifarious applications. It is, in addition, aimed also at other mathematicians, who may themselves work in areas other than the ones being described here. With regard to this latter purpose, I can vouch for its success. For in my own case, I write as a mathematician whose professional interests lie in areas almost entirely outside those described here, and upon reading these articles I have certainly had my perspectives broadened in several ways that I had not expected.

The breadth of the ideas that we find here is considerable, ranging over many areas, such as the philosophy of mathematics, the issue of why mathematics is so important in education and society, and whether its public perception has changed in recent years; perhaps it should now be taught fundamentally differently, and there is the issue of the extent to which the modern technological world might even have thoroughly changed the very nature of our subject. We also find fascinating historical accounts, from achievements made a thousand years or so before the ancient Greeks, to the deep insights and occasional surprising

When we talk about mathematics, or when we teach it, or when we write about it, many of us feign detachment. It is almost a cultural universal to pretend that mathematics is "out there," independent of our whims and oddities. But doing mathematics and talking or writing about it are activities neither neutral nor innocent; we can only do them if we are engaged, and the engagement marks not only us (as thinkers and experimenters) but also those who watch us, listen to us, and think with us. Thus mathematics always requires full participation; without genuine involvement, there is no mathematics.

Mathematicians are also tinkerers—as all innovators are. They try, and often succeed, in creating effective tools that guide our minds in the search for certainties. Or they err; they make judgment mishaps or are seduced by the illusion of infallibility. Sometimes mathematicians detect the errors themselves; occasionally, others point out the problems. In any case, the edifice mathematicians build should be up for scrutiny, either by peers or by outsiders.

And here comes a peculiar aspect that distinguishes mathematics among other intellectual domains: Mathematicians seek validation inside their discipline and community but feel little need (if any) for validation coming from outside. This professional chasm surrounding much of the mathematics profession is inevitable up to a point because of the nature of the discipline. It is a Janus-faced curse of the ivory tower, and it is unfortunate if we ignore it. On the contrary, I believe that we should address it. Seeking the meaning and the palpable reasoning underlying every piece of mathematics, as well as conveying them in natural language, are means that bridge at least part of the gap that separates theoretical mathematics from the general public; such means demythologize the widespread belief that higher mathematics is, by its epistemic status and technical difficulty, inaccessible to the layperson.

Mathematics and its applications are scrutable only as far as mathematicians are explicit with their own assumptions, claims, results, and interpretations. When these elements of openness are missing mathematicians not only fail to disrupt patterns of entrenched thinking but also run the risk of digging themselves new trenches. Writing about mathematics offers freedoms of explanation that complement the dense texture of meaning captured by mathematical symbols.

Talking plainly about mathematics also has inestimable educational and social value. A sign of a mature mind is the ability to hold at the same time opposite ideas and to juggle with them, analyze them, refine them, corroborate them, compromise with them, and choose among them. Such intellectual dexterity has moral and practical consequences, for the lives of the individuals as well as for the life of society. Mathematical thinking is eminently endowed to prepare the mind for these habits, but we almost never pay attention to this aspect, we rarely notice it, and we seldom talk about it. We use contradiction as a trick of the (mathematical) trade and as a routine method of proof. Yet opposing ideas, contrasts, and complementary qualities are intimately interwoven into the texture of mathematics, from definitions and elementary notions to highly specialized mathematical practice; we use them implicitly, tacitly, all the time.

As a reader of this book, you will have a rewarding task in identifying in the contributions some of the virtues I attribute to writing about mathematics—and perhaps many others. With each volume in this series, I put together a book I like to read, the book I did not find in the bookstore years ago. If I include a contribution here, it does not mean that I necessarily concur in the opinions expressed in it. Whether we agree or disagree with other people's views, our polemics gain in substance if we aim to comprehend and address the highest quality of the opposing arguments.

Overview of the Volume

In a sweeping panoramic view of the likely future trajectory of mathematics, Philip Davis asks pertinent questions that illuminate some of the myriad links connecting mathematics to its applications and to other practical domains and offers informed speculations on the multifaceted mathematical imprint on our ever more digitized world.

Ian Stewart explains recent attempts made by mathematicians to refine and reaffirm a theory first formulated by Alan Turing, stating, in its most general formulation, that pattern formation is a consequence of symmetry breaking.

Terence Tao observes that many complex systems seem to be governed by universal principles independent of the laws that govern the interaction between their components; he then surveys various aspects of several well-studied mathematical laws that characterize phenomena as diverse as spectral lines of chemical elements, political elections, celestial mechanics, economic changes, phase transitions of physical materials, the distribution of prime numbers, and others.

A diversity of contexts, with primary focus on social networking, is also Gregory Goth's object of attention; in his article he examines the "small-world" problem—the quest to determine the likelihood that two individuals randomly chosen from a large group know each other.

Charles Seife argues that humans' evolutionary heritage, cultural mores, and acquired preconceptions equip us poorly for expecting and experiencing randomness; yet, in an echo of Tao's contribution, Seife observes that in aggregate, randomness of many independent events does obey immutable mathematical rules.

Writing from experience, Donald Knuth shows that the deliberate and methodical coopting of randomness into creative acts enhances the beauty and the originality of the result.

Soren Johnson discusses the advantages and the pitfalls of using chance in designing games and gives examples illustrative of this perspective.

John Pavlus details the history, the meaning, and the implications of the P versus NP problem that underpins the foundations of computational complexity theory and mentions the many interdisciplinary areas that are connected through it.

Renan Gross analyzes the geometry of the Jerusalem Chords Bridge and relates it to the mathematics used half a century ago by the French mathematical engineer Pierre Bézier in car designs—an elegantly simple subject that has many other applications.

Daniel Silver presents Albrecht Dürer's *Painter's Manual* as a precursor work to projective geometry and astronomy; he details some of the mathematics in the treaty and in Dürer's artworks, as well as the biographical elements that contributed to Dürer's interaction with mathematics.

Kelly Delp writes about the late William Thurston's little-known but intensely absorbing collaboration with the fashion designer Dai Fujiwara and his Issey Mayake team; she describes the topological notions that, surprisingly, turned out to be at the confluence of intellectual passions harbored by two people so different in background and living on opposite sides of the world.

Fiona and William Ross tell the brief history of Jordan's curve theorem, hint at some tricky cases that defy the simplistic intuition behind it, and, most remarkably, illustrate the nonobvious character of the theorem with arresting drawings penned by Fiona Ross.

To answer pressing questions about the need for widespread mathematics education, Anna Sfard sees mathematics as a narrative means to comprehend the world, which we humans developed for our convenience; she follows up on this perspective by arguing that the story we tell (and teach) with mathematics needs to change, in sync with the unprecedented changes of our world.

Erin A. Maloney and Sian L. Beilock examine the practical and psychological consequences of states of anxiety toward mathematical activities. They contend that such feelings appear early in schooling and tend to recur in subsequent years, have a dual cognitive and social basis, and negatively affect cognitive performance. The authors affirm that the negative effect of mathematical anxiety can be alleviated by certain deliberate practices, for instance by writing about the emotions that cause anxiety.

David R. Lloyd reviews arguments put forward by proponents of the idea that the five regular polyhedral shapes were well known, as mathematical objects, perhaps one millennium before Plato—in Scotland, where objects of similar configurations and markings have been discovered. He concludes that the objects, genuine and valuable aesthetically and anthropologically, do not substantiate the revisionist claims at least as far as in the mathematical knowledge they reveal.

In the interaction between the material culture of mathematical instruments and the mathematics that underlined it from the 16th to the 18th centuries in Western Europe, Jim Bennett decodes subtle reciprocal influences that put mathematics in a nodal position of a network of applied sciences, scientific practices, institutional academics, entrepreneurship, and commerce.

Frank Quinn contrasts the main features of mathematics before and after the profound transformations that took place at the core of the discipline roughly around the turn of the 19th century; he contends that the unrecognized magnitude of those changes led to some current fault lines (for instance between teaching and research needs in universities, between school mathematics and higher level mathematics, and others) and in the future might even marginalize mathematics.

Prakash Gorroochurn surveys a collection of chance and statistics problems that confused some of the brilliant mathematical minds of the past few centuries and initially were given erroneous solutions—but had an important historical role in clarifying fine distinctions between the theoretical notions involved.

Elie Ayache makes the logical-philosophical case that the prices reached by traders in the marketplace take precedence over the intellectual speculations intended to justify and to predict them; therefore, the contingency of number prices supersedes the calculus of probabilities meant to model it, not the other way around.

Finally, Kevin Hartnett reports on recent developments related to the *abc* conjecture, a number theory result that, if indeed proven, will have widespread implications for several branches of mathematics.

Other Notable Writings

This section of the introduction is intended for readers who want to read more about mathematics. Most of the recent books I mention here are nontechnical. I offer leads that can easily become paths to research on various aspects of mathematics. The list that follows is not exhaustive, of course—and I omit titles that appear elsewhere in this volume.

Every year I start by mentioning an outstanding recent work; this time is no exception. The reader interested in the multitude of modalities available for conveying data (using graphs, charts, and other visual means) may relish the monumental encyclopedic work *Information Graphics* by Sandra Rendgen and Julius Wiedemann.

An intriguing collection of essays on connections between mathematics and the narrative is *Circles Disturbed*, edited by Apostolos Doxiadis and Barry Mazur. Another collection of essays, more technical but still accessible in part to a general readership, is *Math Unlimited*, edited by R. Sujatha and colleagues; it explores the relationship of mathematics with some of its many applications.

Among the ever more numerous popular books on mathematics I mention Steven Strogatz's *The Joy of X*, Ian Stewart's *In Pursuit of the Unknown*, Dana Mackenzie's *The Universe in Zero Words*, Jeffrey Bennett's *Math for Life*, Lawrence Weinstein's *Guesstimation 2.0*, Norbert Hermann's *The Beauty of Everyday Mathematics*, Keith Devlin's *Introduction to Mathematical Thinking* and Leonard Wapner's *Unexpected Expectations*. Two successful older books that see new editions are *Damned Lies and Statistics* by Joel Best and *News and Numbers* by Victor Cohn and

Lewis Cope. An eminently readable introduction to irrational numbers is appropriately called *The Irrationals,* authored by Julian Havil. A much needed book on mathematics on the wide screen is *Math Goes to the Movies* by Burkard Polster and Marty Ross.

Two venerable philosophers of science and mathematics have their decades-long collections of short pieces republished in anthologies: Hilary Putnam in *Philosophy in an Age of Science* and Philip Kitcher in *Preludes to Pragmatism.* Other recent books in the philosophy of mathematics are *Logic and Knowledge* edited by Carlo Cellucci, Emily Grosholz, and Emiliano Ippoloti; *Introduction to Mathematical Thinking* by Keith Devlin; *From Foundations to the Philosophy of Mathematics* by Joan Roselló; *Geometric Possibility* by Gordon Belot; and *Mathematics and Scientific Representation* by Christopher Pincock. Among many works of broader philosophical scope that take mnemonic inspiration from mathematics, I mention *Spinoza's Geometry of Power* by Valtteri Viljanen, *The Geometry of Desert* by Shelly Kagan, and *The Politics of Logic* by Paul Livingston. Two volumes commemorating past logicians are *Gödel's Way* by Gregori Chaitin and his collaborators, and *Hao Wang* edited by Charles Persons and Montgomery Link. And a compendious third edition of Michael Clark's *Paradoxes from A to Z* has just become available.

In the history of mathematics, a few books focus on particular epochs—for instance, the massive *History of Mathematical Proof in Ancient Traditions* edited by Karine Chemla and the concise *History of the History of Mathematics* edited by Benjamin Wardhaugh; on the works and the biography of important mathematicians, as in *Henri Poincaré* by Jeremy Gray, *Interpreting Newton,* edited by Andrew Janiak and Eric Schliesser, *The King of Infinite Space* [Euclid] by David Berlinski, and *The Cult of Pythagoras* by Alberto Martínez; on branches of mathematics, for example *The Tangled Origins of the Leibnizian Calculus* by Richard Brown, *Calculus and Its Origins* by Richard Perkins, and *Elliptic Tales* by Avner Ash and Robert Gross; or on mathematical word problems, as in *Mathematical Expeditions* by Frank Swetz. An anthology rich in examples of popular writings on mathematics chosen from several centuries is *Wealth in Numbers* edited by Benjamin Wardhaugh. Two collections of insightful historical episodes are Israel Kleiner's *Excursions in the History of Mathematics* and Alexander Ostermann's *Geometry by Its History.* And an intriguing attempt to induce mathematical rigor in historical-religious controversies is *Proving History* by Richard Carrier.

The books on mathematics education published every year are too many to attempt a comprehensive survey; I only mention the few that came to my attention. In the excellent series *Developing Essential Understanding* of the National Council of Teachers of Mathematics, two recent substantial brochures written by Nathalie Sinclair, David Pimm, and Melanie Skelin focus on middle school and high school geometry. Also at the NCTM are *Strength in Numbers* by Ilana Horn; an anthology of articles previously published in *Mathematics Teacher* edited by Sarah Kasten and Jill Newton; and *Teaching Mathematics for Social Justice,* edited by Anita Wager and David Stinson. Other books on equity are *Building Mathematics Learning Communities* by Erika Walker and *Towards Equity in Mathematics Education* edited by Helen Forgasz and Ferdinand Rivera. In a niche of self-help books I would include Danica McKellar's *Girls Get Curves* and Colin Pask's *Math for the Frightened.* A detailed ethnomathematics study is Geoffrey Saxe's *Cultural Development of Mathematical Ideas* [in Papua New Guinea]. And a second volume on *The Mathematics Education of Teachers* was recently issued jointly by the American Mathematical Society and the Mathematical Association of America.

In a group I would loosely call applications of mathematics and connections with other disciplines, notable are *Fractal Architecture* by James Harris, *Mathematical Excursions to the World's Great Buildings* by Alexander Hahn, *Proving Darwin* by Gregory Chaitin, *Visualizing Time* by Graham Wills, *Evolution by the Numbers* by James Wynn, and *Mathematics and Modern Art* edited by Claude Bruter. Slightly technical but still widely accessible are *Introduction to Mathematical Sociology* by Phillip Bonacich and Philip Lu, *The Essentials of Statistics* [for social research] by Joseph Healey, and *Ways of Thinking, Ways of Seeing* edited by Chris Bissell and Chris Dillon. More technical are *The Science of Cities and Regions* by Alan Wills, and *Optimization* by Jan Brinkhuis and Vladimir Tikhomirov.

I conclude by suggesting a few interesting websites. "Videos about numbers & stuff" is the deceptively self-deprecating subtitle of the Numberphile (http://www.numberphile.com/) page, hosting many instructive short videos on simple and not-so-simple mathematical topics. A blog that keeps current with technology that helps teaching mathematics is Mathematics and Multimedia (http://mathandmultimedia.com/); also a well-done educational site is Enrich Mathematics (http://nrich.maths.org/frontpage) hosted by Cambridge University.

Two popular sites for exchanging ideas and asking and answering questions are the Math Overflow (http://mathoverflow.net/) and the Mathematics StackExchange (http://math.stackexchange.com/). The Parameterized Complexity (http://fpt.wikidot.com/) page offers a useful forum of information and instruction for those interested in mathematical connections to computing, psychology, and cognitive sciences. A brief history of mathematics is available on the Story of Mathematics (http://www.storyofmathematics.com/index.html) site. Many institutions host a wealth of materials on their internet sites; I mention here the Harvey Mudd College (http://www.math.hmc.edu/funfacts/), the Clay Mathematics Institute (http://www.claymath.org/index.php), and the Cornell Mathematics Library (http://mathematics.library.cornell.edu/).

I hope you, the reader, find the same value and excitement in reading the texts in this volume as I found while searching, reading, and selecting them. For comments on this book and to suggest materials for consideration in preparing future volumes, I encourage you to send correspondence to me at Mircea Pitici, P.O. Box 4671, Ithaca, NY 14852.

Works Mentioned

Ash, Avner, and Robert Gross. *Elliptic Tales: Curves, Counting, and Number Theory*. Princeton, NJ: Princeton University Press, 2012.

Barnsley, Michael F. *Fractals Everywhere*. Mineola, NY: Dover, 2012.

Belot, Gordon. *Geometric Possibility*. Oxford, UK: Oxford University Press, 2011.

Bennett, Jeffrey. *Math for Life: Crucial Ideas You Didn't Learn in School*. Greenwood Village, CO: Roberts & Co. Publishers, 2012.

Berlinski, David. *The King of Infinite Space: Euclid and His Elements*. New York: Basic Books, 2012.

Best, Joel. *Damned Lies and Statistics: Untangling Numbers from the Media, Politicians, and Activists*, updated edition. Berkeley, CA: University of California Press, 2012.

Bissell, Chris, and Chris Dillon. (Eds.) *Ways of Thinking, Ways of Seeing: Mathematical and Other Modeling in Engineering and Technology*. Heidelberg, Germany: Springer Verlag, 2012.

Bonacich, Phillip, and Philip Lu. *Introduction to Mathematical Biology*. Princeton, NJ: Princeton University Press, 2012.

Brinkhuis, Jan, and Vladimir Tikhomirov. *Optimization: Insights and Applications*. Princeton, NJ: Princeton University Press, 2012.

Brown, Richard C. *The Tangled Origins of Leibnizian Calculus: A Case Study of Mathematical Revolution*. Singapore: World Scientific, 2012.

Bruter, Claude. (Ed.) *Mathematics and Modern Art*. New York: Springer-Verlag, 2012.

Carrier, Richard C. *Proving History: Bayes's Theorem and the Quest for the Historical Jesus*. Amherst, MA: Prometheus Books, 2012.

Wills, Graham. *Visualizing Time*. Heidelberg, Germany: Springer Verlag, 2012.
Wilson, Alan. *The Science of Cities and Regions: Lectures on Mathematical Model Design*. New York: Springer Science+Business Media, 2012.
Wynn, James. *Evolution by the Numbers: The Origins of Mathematical Argument in Biology*. Anderson, SC: Parlor Press, 2012.

The **BEST** **WRITING** on **MATHEMATICS**

2013

The Prospects for Mathematics in a Multimedia Civilization

PHILIP J. DAVIS

I. Multimedia Mathematics

First let me explain my use of the phrase "multimedia civilization."* I mean it in two senses. In my first usage, it is simply a synonym for our contemporary digital world, our click-click world, our "press 1,2, or 3 world", a world with a diminishing number of flesh-and-blood servers to talk to. This is our world, now and for the indefinite future. It is a world that in some tiny measure most of us have helped make and foster.

In my second usage, I refer to the widespread and increasing use of computers, fax, e-mail, the Internet, CD-ROMs, iPods, search engines, PowerPoint, and YouTube—in all mixtures. I mean the phrase to designate the cyberworld that embraces such terms as interface design, cybercash, cyberlaw, virtual-reality games, assisted learning , virtual medical procedures, cyberfeminism, teleimmersion, interactive literature, cinema, and animation, 3D conferencing, and spam, as well as certain nasty excrescences that are excused by the term "unforeseeable developments." The word (and combining form) "cyber" was introduced in the late 1940s by Norbert Wiener in the sense of feedback and control. Searching on the prefix "cyber" resulted in 304,000,000 hits, which, paradoxically, strikes me as a lack of control.

I personally cannot do without my word processor, my mathematical software, and yes, I must admit it, my search engines. I find I can check conjectures quickly and find phenomena accidentally. (It is also

*This article is an expanded and newly updated version of a Urania Theater talk given as part of the International Congress of Mathematicians, Berlin, Germany, Aug. 21, 1998.

the case that I find too many trivialities!) As a writer, these tools are now indispensable for me.

Yes, the computer and all its ancillary spinoffs have become a medium, a universal forum, a method of communication, an aid both to productive work and to trouble making, from which none of us are able to escape. A mathematical engine, the computer is no longer the exclusive property of a few mathematicians and electrical engineers, as it was in the days of ENIAC *et alia* (late 1940s). Soon we will not be able to read anything or do any "brain work" without a screen in our lap and a mouse in our hands. And these tools, it is said, will soon be replaced by Google glasses and possibly a hyperGoogle brain. We have been seduced, we have become addicts, we have benefited, and we hardly recognize or care to admit that there is a downside.

What aspects of mathematics immersed in our cyberworld shall I consider? The logical chains from abstract hypotheses to conclusions? Other means of arriving at mathematical conclusions and suggesting actions? The semiotics of mathematics? Its applications (even to multimedia itself!)? The psychology of mathematical creation? The manner in which mathematics is done; is linked with itself and with other disciplines; is published, transmitted, disseminated, discussed, taught, supported financially, and applied? What will the job market be for its young practitioners? What will be the public's understanding and appreciation of mathematics? Ideally, I should like to consider all of these. But, of course, every topic that I've mentioned would deserve a week or more of special conferences and would result in a large book.

Poincaré's Predictions

We have now stepped into the new millennium, and inevitably this step suggests that I project forward in time. Although such projections, made in the past, have proved notoriously inadequate, I would be neglecting my duty if I did not make projections, even though it is guaranteed that they will become the objects of future humorous remarks.

Here's an example from the past. A century ago, at the Fourth International Congress of Mathematicians held at Rome in 1908, Henri Poincaré undertook such a task. In a talk entitled "*The Future of Mathematics,*" Poincaré mentioned 10 general areas of research and some specific problems within them, which he hoped the future would resolve.

What strikes me now in reading his article is not the degree to which these areas have been developed—some have—but the inevitable omission of a multiplicity of areas that we now take for granted and that were then only *in utero* or not even conceived. Though the historian can always find the seeds of the present in the past, particularly in the thoughts of a mathematician as great as Poincaré, I might mention as omissions from Poincaré's prescriptive vision the intensification of the abstracting, generalizing, and structural tendencies; the developments in logic and set theory; pattern theory; and the emergence of new mathematics attendant upon the physics of communication theory, fluids, materials, relativity, quantum theory, and cosmology. And of course, the computer, in both its practical and theoretical aspects; the computer, which I believe is the most significant mathematical development of the 20th century; the computer, which has altered our lives almost as much as the "infernal" combustion engine and which may ultimately surpass it in influence.

Poincaré's omission of all problems relating immediately to the exterior world—with the sole exception (!) of Hill's theory of lunar motion—is also striking.

How then should the predictor with a clouded vision and limited experience proceed? Usually by extrapolating forward from current tendencies that are obvious even to the most imperceptive observer.

What Will Pull Mathematics into the Future?

Mathematics grows from external pressures and from pressures internal to itself. I think the balance will definitely shift away from the internal and that there will be an increased emphasis on applications. Mathematicians require support; why should society support their activity? For the sake of pure art or knowledge? Alas, we are not classic Greeks, who scorned those who needed to profit from what they learned, or 18th century aristocrats, for whom science was a hobby. And even the material generated by these groups was pulled along by astronomy and astrology (for which a charge was made), geography, navigation, and mechanics. Society will now support mathematics generously only if it promises bottom-line benefits.

Now focus on the word "benefits." What is a benefit? Richard Hamming of the old Bell Telephone Laboratories said in a famous epigraph to his book on scientific computation, “The object of computation is not numbers but insight.” Insight into a variety of physical and social processes, of course. But I perceive (40 years after Hamming’s book and with a somewhat cynical eye) that the real object of computation, commercial and otherwise, is often neither numbers nor insight nor solutions to pressing problems, but worked on by physicists and mathematicians, to perfect money-making products. Often computer usages are then authorized by project managers who have little technical knowledge. The Descartesian precept *cogito ergo sum* has been replaced by *producto ergo sum.*

If, by chance, humanity benefits from this activity, then so much the better; everybody is happy. And if humanity suffers, the neo-Luddites will cry out and form chat groups on the Web, or the hackers will attack computer systems or humans. The techno-utopians will explain that you can’t make omelets without breaking a few eggs. And pure mathematicians will follow along, moving closer to applications while justifying the purity of their pursuits to the administrators, politicians, and the public with considerable truth that one never knows in advance what products of pure imagination can be turned to society’s benefit. The application of the theory of numbers to cryptography and the (Johann) Radon transform and its application to tomography have been displayed as shining examples of this. Using that most weasel of rhetorical expressions, “in principle,” in principle, all mathematics is potentially useful.

I could use all my space describing many applications that seem now to be hot and are growing hotter. I will mention several and comment briefly on but a few of them. In selecting these few, I have ignored “pure” fields out of personal incompetence. I simply do not have the knowledge or authority to single out from a hundred expanding subfields the ones with particularly significant potential and how they have fared via multimedia. For more comprehensive and authoritative presentations, I recommend *Mathematics: Frontiers and Perspectives* and *Mathematics Unlimited—2001 and Beyond.*

Mathematics and the Physical and Engineering Sciences

These have been around since Galileo, but Newton’s work was the great breakthrough. However, only in the past hundred years or so has theoretical mathematics been of any great use to technology. The pursuit of physical and engineering sciences is today unthinkable without significant computational power. The practice of aerodynamic design has altered significantly, but the "digital wind tunnel" has not yet arrived, and some have said it may never. Theories of turbulence are not yet in satisfactory shape—how to deal with widely differing simultaneous scales continues to perplex. Newtonians who deal with differential-integral systems must learn to share the stage with a host of probabilists with their stochastic equations. Withal, hurricane and tornado predictions continue to improve, perhaps more because of improvements in hardware (e.g., real-time data from aircraft or sondes and from nanocomputers) than to the numerical algorithms used to deal with the numerous models that are in use. Predictions of earthquakes or of global warming are controversial and need work. Wavelet, chaos, and fractal theorists and multiresolution analysts are hard at work hoping to improve their predictions.

Mathematics and the Life Sciences

Mathematical biology and medicine are booming. There are automatic diagnoses. There are many models around in computational biology; most are untested. One of my old Ph.D. students has worked in biomolecular mathematics and designer drugs. He and numerous others are now attempting to model strokes via differential equations. Good luck!

I visited a large hospital recently and was struck by the extent that the aisles were absolutely clogged with specialized computers. Later, as a patient, I was all wired up and plugged into such equipment with discrete data and continuous waveforms displayed bedside and at the nurses' stations. Many areas of medical and psychological practice have gone or are going virtual. There is no doubt that we are now our own digital avatars and we are all living longer and healthier lives. In this development, mathematics, though way in the background and though not really understood by the resident physicians or nurses, has played a significant role.

Work on determining and analyzing the human genome sequences, with a variety of goals in mind and using essentially combinatorial and probabilistic methods, is a hot field. In the past decade, the cost of DNA sequencing has come down dramatically.

Genetic engineering on crops goes forward but has raised hackles and doomsday scenarios.

are class action and discrimination suits based on statistical evidence. Multiple regression enters into the picture strongly. Mathematical algorithms themselves have been scrutinized and may be subject to litigation as part of intellectual property. In the burgeoning field of "jurimath" or "jurimetrics," there are now texts for lawyers and a number of journals. This field should be added to the roster of applications of mathematics and should be taught in colleges and law schools.

Consider polls. We spend millions and millions of dollars polling voters, polling consumers, asking people how they feel about anything at all. Consider the census. How should one count? Counting, the simplest, most basic of all mathematical operations, in the sharp theoretical sense, turns out to be a practical impossibility. Sampling is recommended; it reduces the variance but increases the discrepancy. It has been conjectured that sampling will increase the power of the minority party, hence the majority party is against it. As early as Nov. 30, 1998, the case was argued before the U.S. Supreme Court. Despite all these developments, we are far from Leibniz' dream of settling human disputes by computation.

Mathematics in the Service of Cross- or Trans-Media

Here are a few instances:

Music ⇔ score
Oral or audio ⇔ hard copy

Motion has been captured from output from a wired-up human, then analyzed and synthesized. (The animation of Gollum in *Lord of the Rings* was produced in this way.)
Balletic motion ⇔ Choreographic notation

Voice → action. Following the instructions of an electronic voice, I push a few buttons and a ticket gets printed out, all paid for. But the buttons are becoming fewer as voice interpretation improves daily.

Under this rubric, I would also place advanced search methods that go beyond the "simple" Google type. Image recognition and feature identification, often using statistics, lead to the lively field of searching for image or audio content (e.g., find me a Vermont landscape with a red barn and a large pile of pumpkins). IDs and person identification schemes are products here.

Mathematics and Education

Though computer science departments declared their independence of mathematics or electrical engineering departments in the mid-'70's, undergraduate majors in computer science are usually required to have a semester or two of calculus and a semester of discrete math. Depending on what kind of work they intend to do, it may be useful for them to have additionally linear algebra, probability, geometry, numerical analysis, or mathematical logic. Thus, mathematics or the spirit of mathematics, though it may be hidden from view, lurks behind most theoretical or commercial developments in the field.

Regarding the changes in classroom education, one of my colleagues wrote me as follows:

> My teaching has already changed a great deal. Assignments, etc. go on the web page. Students use e-mail to ask questions which I then bring up in class. They find information for their papers out there on the web. We spend one day a week doing pretty serious computing, producing wonderful graphics, setting up the mathematical part of it and dumping the whole mess into documents that can be placed on a web page. I am having more fun than I used to, and the students appear to be having a pretty good time while learning a lot. Can all this be bad?

A distinguished applied mathematician of my acquaintance is spending part of his time producing CD-ROMs to publicize his theories and experiences. The classic modes of elementary and advanced teaching have been amplified and sometimes displaced by computer products. Thousands of lectures, often illustrated and interactive, on all conceivable mathematical topics, can be downloaded. Automated correction of papers has been going on for decades. There is software to detect plagiarism.

A good university computer store has more of these products for sale than there are brands of cheese in the gourmet market. Correspondence courses via such things as e-mail are flourishing. "Web universities" that have up to 100,000 students per class have sprung up all over the world. Will the vast armies of flesh-and-blood teachers become obsolete?

Working Habits and the Working Environment

A colleague wrote to me

> On balance, I believe that science will suffer in the multi-media age. My experience is that true thinking now goes against the grain. I feel I have to be rude to arrange for a few peaceful hours a day for real work. Saying no to too many invitations, writing short answers to too many e-mail questions about research. A letter comes to me from a far corner of the world: "Please explain line six of your 1987 paper." I stay home in the mornings hiding from my office equipment. Big science projects, interdisciplinary projects, big pushes, aided and abetted by multi-media and easy transportation have diminished my available time for real thought. I am also human and succumb to the glamour of today's technoglitz.

Another colleague said of his mathematical research experiences,

> Publication on line is easy. But there is too much of it, often relating trivial advances. What is out there often lacks depth and goes unrefereed.

The book world, or more generally the world of information production, dissemination, credit, and copyright royalties is in an absolute turmoil. Is all this really for the better? *Qui vivra verra.*

Dissemination

A half century ago, mathematicians used to relegate certain jobs to other disciplines or crafts or professions. We have now become our own typists, typesetters, drafters, library scientists, book designers, publishers, jobbers, public relations agents, and sales agents. For all that the computer is rapid, these activities absorb substantial blocks of time and energy that were formerly devoted to thinking about problems.

Occasionally, in the past, scientists had to leave pure creation and discovery and worry about dissemination. In 1597, when Tycho Brahe moved from Copenhagen to Prague to take a job there, he lugged along his own heavy printing press. Journals? Who needs them now when you can download papers by the hundreds? The dissemination of mathematics through textbooks and learned journals is threatened with obsolescence in favor of online electronic publishing. Every man and woman is now a journal "publisher" or a mathematical blogger. I exchange PDF files with a colleague 7,000 miles away.

Whereas people such as Copernicus and Newton waited years before they published, today's scientists, under a variety of pressures, go on line with electronic publishing before their ideas are out of the oven, often unchecked; and they can receive equally rapid and equally half-baked feedback. Refereeing has diminished. The authority that once attached to the published page has vanished.

Are books obsolete? A collaborator working with me and much in love with mathematical databases, picked up (even after advanced filtering) more than 100,000 references to a key word that was relevant to our work. This information overdose produced an immediate blockage or atrophy of the spirit in him. How can we afford the time to assess this raw, unassimilated information overload ? Should we then simply plow ahead on our own as best we can and hope that we will come up with something new and of interest ? Semioticist and novelist Umberto Eco wrote, in *How to Travel with a Salmon,* ". . . the whole information industry runs the risk of no longer communicating anything because it tells too much."

My eyebrows were raised recently when I learned that as part of a large grant application to the National Science Foundation, the applicants were advised to include a detailed plan for the dissemination of their work. In the multimedia age, mathematics is being transformed into a product to be marketed as other products. Just as, for a price, a person seeking employment can get a company to produce a fancy CV, there are or there will be, I feel sure, companies that, for a price, will undertake to subcontract the dissemination of mathematical applications.

The Public Understanding of Mathematics

Every aspect of our lives is increasingly being mathematized. We are dominated by and we are accommodating to mathematical theories, engines, and the arrangements they prescribe. Yet, paradoxically, the nature of technology makes it possible, through chipification, for the mathematics itself to disappear into the background and for the public to be totally unaware of it. It is probably the case that despite the claims of teachers and educational administrators, the general population

Despite Einstein's quip that God doesn't play dice with the universe, the relevance of probability theory is now strong and becoming stronger. Niels Bohr rapped Einstein's knuckles: "Albert, don't tell God what he should do with his dice." David Mumford opines that probability theory and statistical reasoning

> will emerge as better foundations for scientific models . . . and as essential ingredients of theoretical mathematics, even the foundations of mathematics itself.

Thinking vs. Clicking

I have heard over and over again from observers that thinking increasingly goes against the grain. Is thinking obsolete or becoming more obsolete? To think is to click. To click is to think. Are these the equations for the future? Did not mathematician and philosopher Alfred North Whitehead write in one of his books that it was a mistake to believe that one had constantly to think? Did not Descartes write that his specific goal was to bring about a condition of automated reasoning? Do not the rules, the paradigms, the recipes, the algorithms, the theorems, and the generalizations of mathematics reduce once again the necessity for thinking through a certain situation?

Experimental mathematics and visual theorems, all linked to computer experiences, are increasing in frequency. There are now two types of researchers: the first try to think before they compute; others do the reverse. I cannot set relative values on these strategies, but it is clear that the successful development of mathematics has in the past has been enriched by the simultaneous use of both strategies.

A researcher in automatic intelligence (AI) has written me,

> Your question "Is Thinking Obsolete" is very much to the point. This has certainly been the trend in AI over to the past ten years (just now beginning to reverse itself)—trying to accomplish things through huge brute-force searches and statistical analyses rather than through high-level reasoning.

We can also ask in this context, "Are certain parts of traditional advanced mathematics obsolete?" For example, what portions of a theory of differential equations retain value when numerical solutions are available on demand, and when, for many equations, computation is far ahead of analytic or theorematic mathematics in explaining what is going on? Yet, theory points directions in which we should look experimentally; otherwise, we can wander at random fruitlessly.

Words or Mathematical Symbols vs. Icons

A semanticist, Mihai Nadin, has written a large book, *The Civilization of Illiteracy,* on the contemporary decline of the printed word, how the word is being displaced by the hieroglyphic or iconic modes of communication. There is no doubt in my mind but that such a displacement will have a profound effect on the inner texture of mathematics. Such a shift already happened 4,000 years ago. Numbers are among the oldest achievements of civilization, predating, perhaps, writing. In his famous book, *Über Vorgreichischer Mathematik*, Otto Neugebauer "explains . . . how hieroglyphs and cuneiforms are written and how this affects the forms of numbers and the operations with numbers." Another such shift occurred in the late Middle Ages, when, with strong initial resistance, algebraic symbolisms began to invade older texts.

Mathematics as Objective Description vs. Mathematics by Fiat, or the Ideal vs. the Constructed and the Virtual

Applied mathematics deals with descriptions, predictions, and prescriptions. We are now in a sellers' market for all three. Prescriptions will boom. There may indeed be limits to what can be achieved by mathematics and science (there are a number of books on this topic), but I see no limits to the number of mathematizations that can be prescribed and to which humans are asked to conform. In the current advanced state of the mathematization of society and human affairs, we prescribe the systems we want to put in, from the supermarket to the library, to the income tax, to stocks and bonds, to machines in the medical examination rooms. All products, all human activities are now wide open to prescriptive mathematizations. The potentialities and the advantages envisaged and grasped by the corporate world will lead it to pick up some of the developmental tab. And, as it does, the human foot will be asked, as with Cinderella's sisters, to fit the mathematical shoe. And if the shoe does not fit, tough for the foot.

What Is Proved vs. What Is Observed

This is the philosophical argument between Descartes and Giambattista Vico. I venture that as regards the generality of makers and users of mathematics, its proof aspect will diminish. Remember, mathematics does not and never did belong exclusively to those who happen to call themselves mathematicians and who pursue a rather rigid notion of mathematics. I would hope that the notion of proof will be expanded so as to be acknowledged as simply one part of a larger notion of "mathematical evidence."

The whole present corpus of mathematical experience and education has come under attack from at least two different sociopolitical directions: European or Western mathematics vs. other national or ethnic mathematics. We have today's ethnomathematicians to thank for reminding us that different cultures, primitive and advanced, have had different answers as to what mathematics is and how it should be pursued and valued (e.g., ancient oriental mathematics was carried on in a proof-free manner, and ancient Indian mathematics often expressed itself in verse.) More important than drawing on ancient, "non-Western" material is the possibility that new "ethnic" splits, to be described momentarily, will emerge from within current practices. Will a civilization of computer-induced illiteracy compel major paradigm shifts in mathematics? Extrapolating from Nadin's book, one might conclude that this might arrive sooner than we think and perhaps more rapidly than is good for us.

Male vs. Female Mathematics?

Mathematics has been perceived as an expression of male machismo. Margaret Wertheim is a TV writer, a promoter of visual mathematics, and a former student of math and physics. Let me quote from her book *Pythagoras' Trousers*:

> One of the reasons more women do not go into physics is that they find the present culture of this science and its almost antihuman

> focus, deeply alienating. . . . After six years of studying physics and math at university, I realized that much as I loved the science itself, I could not continue to operate within such an intellectual environment. (p 15)

The bottom line of this *Pythagoras' Trousers* is that if more women were in mathematics and science (particularly in physics), then they would create

> an environment in which one could pursue the quest for mathematical relationships in the world around us, but within a more human ethos. . . . The issue is not that physics is done by men, but rather the kind of men who have tended to dominate it. . . . Mathematical Man's problem is neither his math nor his maleness per se, but rather the pseudo religious ideals and self-image with which he so easily becomes obsessed.

In point of fact, more women are entering mathematics and science, and it will take at least several generations to observe whether or not Wertheim's vision will materialize.

The Apparent vs. the Occult

In a somewhat disturbing direction, some mathematicians and physicists have been producing hermeticisms, apocalypses of various sorts, final theories of everything, secret messages hidden in the Bible, everything under the sun implied by Gödel's theorem. I was shocked recently to read that one of the mathematical societies in the United States had published some of this kind of material—even though it was in a spirit of "fun." The old marriage of literacy and rationality, in place since the Western Enlightenment, seems to be cracking a bit. Rationality can shack up with fanaticisms. Is this part of a reaction of a mathematized civilization with its claims to logical rigidity?

Soft Mathematics vs. Traditional Mathematics

I have picked up the term "soft mathematics" from Keith Devlin's popular book *Goodbye, Descartes,* which describes the difficulties of the relationship among natural language, logic, and rationality. These difficulties, Devlin asserts, cannot be overcome by traditional mathematics of the Cartesian variety, and he hopes for the development of a "soft mathematics"— not yet in existence—that "will involve a mixture of mathematical reasoning, and the less mathematically formal kinds of reasoning used in the social sciences." Devlin adds that, "perhaps most of today's mathematicians find it hard to accept the current work on soft mathematics as 'mathematics' at all." Nonetheless, some see the development as inevitable, and Devlin uses as a credentialing authority the mathematician-philosopher Gian-Carlo Rota. Rota comes to a similar viewpoint through his phenomenological (e.g., Husserl and Heidegger) orientation. After listing seven properties that phenomenologists believe are shared by mathematics (absolute truth; items, not objects; nonexistence; identity; placelessness; novelty; and rigor), Rota goes on to say,

> Is it true that mathematics is at present the only existing discipline that meets these requirements? Is it not conceivable that someday, other new, altogether different theoretical sciences might come into being that will share the same properties while being distinct from mathematics?

Rota shares Husserl's belief that a new Galilean revolution will come about to create an alternative, soft mathematics, that will establish theoretical laws through idealizations that run counter to common sense.

And what is "common sense?" It may be closer than we think to what George Bernard Shaw wrote in *Androcles and the Lion,* "People believe not necessarily because something is true but because in some mysterious way it catches their imagination." The vaunted, metaphysical and (I think) mythic unity of mathematics is further threatened by self-contained, self-publishing chat groups. It was already threatened in Poincaré's day by the sheer size of the material available. The riches of mathematics, without contemplative judgments, would, in the words of Poincaré, "soon become an encumbrance and their increase produce an accumulation as incomprehensible as all the unknown truths are to those who are ignorant."

The classic Euclidean mode of exposition and teaching, "definition, theorem, proof," has come under serious attack as not providing a realistic description of how mathematics is created, grasped, or used. Platonism and its various offspring, which have been the generally accepted philosophies of mathematicians, have come under serious attack. Here are a few quotes that bear on this:

> By giving mathematicians access to results they would never have achieved on their own, computers call into question the idea of a transcendental mathematical realm. They make it harder and harder to insist as the Platonists do, that the heavenly content of mathematics is somehow divorced from the earthbound methods by which mathematicians investigate it. I would argue that the earthbound realm of mathematics is the only one there is. And if that is the case, mathematicians will have to change the way they think about what they do. They will have to change the way they justify it, formulate it and do it.
>
> —Brian Rotman

> I know that the great Hilbert said "We will not be driven out of the paradise that Cantor has created for us." And I reply: "I see no need for walking in."
>
> —Richard Hamming

> I think the Platonistic philosophy of mathematics that is currently claimed to justify set theory and mathematics more generally is thoroughly unsatisfactory, and that some other philosophy grounded in inter-subjective human conceptions will have to be sought to explain the apparent objectivity of mathematics.
>
> —Solomon Feferman

> In the end it wasn't Gödel, it wasn't Turing and it wasn't my results that are making mathematics go in an experimental direction. The reason that mathematicians are changing their habits is the computer.
>
> —G. J. Chaitin

A number of philosophies, though buckets of ink are still spilled by their adherents, have suffered from "dead-end-itis." Is philosophy, in general, irrelevant in today's world? But philosophy will never disappear for formulated or unformulated thoughts; it is what we believe deep down to be the nature of things. And we shall have more of it before we can say *nunc dimittis*.

A Philosophy of Mathematics Fostered by Multimedia

Philosopher George Berkeley (1685–1753) has resurfaced as the philosopher of choice for virtual reality. *"Esse est percipi"* (to be is to be perceived, and vice versa) is Berkeley's tag line (often refuted). Cynthia M. Grund is a mathematician, a philosopher, and a Scholar in Residence, 2007–2008, at Whitehall, Berkeley's temporary residence in Middletown, Rhode Island. Using a virtual reality package known as "Second Life," she has immersed herself as avatar in a computer construction of Whitehall:

> By juxtaposing features borrowed directly from the original Whitehall with novel features only available in a virtual environment, but inspired by aspects of the original one, the project explores and exemplifies the philosophical relationship between real and virtual in an educational context, as well as philosophical aspects of issues such as the relationship of user to avatar, relationships among avatars, the relationship of avatars to "their" world, to name only a few.

The experience of negotiating one's virtual self though a 3-D, virtual metaworld raises questions as to the relation among sight, touch, and the other senses and of the relation between the perceiver and the perceived. These questions have been discussed along neo-Berkeleyan lines.

III. A Personal Illumination

Here, then, are some of the "tensions of mathematical texture" that I perceive. Today's scientist or mathematician spends his or her days in a way that is vastly different from 50 years ago, even 20 years ago. Thinking now is accomplished differently. Science is now undergoing a fundamental change; it may suffer in some respects, but it will certainly create its own brave new world and proclaim new insights and idealisms. I think there will be a widening to what has been traditionally been considered to be valid mathematics. In the wake of this, the field will again be split just as it was in the late 1700s, when it began to be split into the pure and the applied. As a consequence, there will be the "true believers," pursuing the subject pretty much in the traditional manner, and the radical wave of "young Turks," pursuing it in ways that will raise the eyebrows and the hackles of those who will cry, "They are traitors to the great traditions."

In his autobiography, Elias Canetti, Nobelist in literature (1981), speaks of an illumination he had as a young man. Walking along the streets of Vienna, he saw in a flash that history could be explained by the tension between the individual and the masses. Yes, we may consider how mathematics in our multimedia age has affected separately both the individual and society. But walking the streets of my hometown, I had my own illumination: that the history of future mathematics will be seen as the increased tension and increased interfusion, sometimes productive, sometimes counterproductive, between the real and the virtual. What meaning the future will give to these last two adjectives and how these elements will play out are now most excellent questions to put to the mathematical oracles.

Thanks

I thank Robert Barnhill, Fred Bisshopp, Bernhelm Booss-Bavnbek, Ernest Davis, John Ewing, Stuart Geman, David Gottlieb, Cynthia M. Grund, John Guckenheimer, Arieh Iserles, David Mumford, Igor Najfeld, and Glen Pate.

Bibliography

V. Arnold, M. Atiyah, P. Lax, B. Mazur, eds., *Mathematics: Frontiers and Perspectives*. American Mathematical Society, 2000.

Nicolaus Bernoulli, *"De Usu Artis Conjectandi in Jure," Dissertatio Inauguralis*. Basel, 1709. Reprinted in Jakob Bernoulli, *Werke*. vol. 3.

Bernhelm Booss-Bavnbek and Jens Høyrup, eds., *Mathematics and War*. Birkhauser, 2003.

G. J. Chaitin, *The Limits of Mathematics*. Springer Verlag, 1997.

Philip J. Davis, "Ein Blick in die Zukunft: Mathematik in einer Multi-media-Zivilisation," in: *Alles Mathematik*, M. Aigner and E. Behrends, eds., Vieweg-Teubner, 2000.

Philip J. Davis, *Mathematics and Common Sense*. A. K. Peters, 2007.

Philip J. Davis, "Unity and Disunity in Mathematics," *European Math. Soc. Newsletter*, March 2013.

Philip J. Davis and David Mumford, "Henri's Crystal Ball," *Notices of the AMS*, v. 55, No. 4, April 2008, pp. 458–466.

Keith Devlin, *Goodbye, Descartes: The End of Logic and the Search for a New Cosmology of the Mind*. John Wiley, New York, 1997.

Umberto Eco, *How to Travel with a Salmon and Other Essays*. Harcourt Brace, 1994.

Björn Engquist and Wilfried Schmidt, *Mathematics Unlimited—2001 and Beyond*. Springer, 2001.

P. Etingof, V. Retach, I. M. Singer, eds., *The Unity of Mathematics*. Birkhauser, 2006.
William Everdell, *The First Moderns*. Univ. Chicago Press, 1996.
Hans Magnus Enzensberger, *Critical Essays*. New York, Continuum, 1982, esp. "The Industrialization of the Mind," pp. 3–14.
Solomon Feferman,"Does mathematics need new axioms ?" 1999, available on the Internet.
Michael O. Finkelstein and Bruce Levin, *Statistics for Lawyers*. Springer Verlag, 1990.
Jeremy Gray, *Henri Poincaré: A Scientific Biography*. Princeton Univ. Press, 2013.
Cynthia M. Grund, *Perception and Reality in – and out – of* Second Life: Second Life *as a Tool for Reflection and Instruction at the University of Denmark,* available at http://eunis.dk/papers/p10.pdf (last visited Apr. 17, 2013).
Richard W. Hamming, *Introduction to Applied Numerical Analysis*. McGraw-Hill, 1971.
Richard W. Hamming, "Mathematics on a Distant Planet," *American Mathematical Monthly*, vol. 105, no. 7, Aug.–Sept. 1998, pp. 640–650.
Reuben Hersh, "Some Proposals for Reviving the Philosophy of Mathematics," *Advances in Mathematics*, v. 31, 1979. pp. 31–50. Reprinted in Tymoczko, Thomas, *In the Philosophy of Mathematics*. Princeton Univ. Press, Princeton, N.J., pp. 9–28.
Reuben Hersh, *What Is Mathematics, Really?* Oxford U. Press, 1997.
David Mumford, *Trends in the Profession of Mathematics*. Presidential Address, International Mathematical Union, Jahresbuch der Deutsche Mathematischer Vereinigung (DMV), 1998.
Mihai Nadin, *The Civilization of Illiteracy*. Dresden Univ. Press, 1997.
Otto Neugebauer, *Über Vorgreichischer Mathematik*. Leipzig, 1929.
Henri Poincaré, *The Future of Mathematics*. (Address delivered at International Congress of Mathematicians, 1908, Rome.) Translated and reprinted in Annual Report of the Smithsonian Institution, 1909, pp. 123–140. Available on the Internet.
Gian-Carlo Rota, *Ten Remarks on Husserl and Phenomenology*. Address delivered at the Provost's Seminar, MIT, 1998.
Brian Rotman, "*The Truth about Counting,*" *The Sciences,* Nov.–Dec. 1977.
Brian Rotman, *Ad Infinitum: The Ghost in Turing's Machine: Taking God out of Mathematics and Putting the Body Back In*. Stanford Univ. Press, 1993.
T. Tymoczko,ed., *New Directions in the Philosophy of Mathematics*. Birkhauser, 1986.
Wertheim, Margaret. (1995). *Pythagoras' Trousers*. New York: Crown.

Fearful Symmetry

Ian Stewart

> Tyger! Tyger! burning bright
> In the forests of the night,
> What immortal hand or eye
> Could frame thy fearful symmetry?

In this opening verse of William Blake's "The Tyger" from his *Songs of Experience* of 1794, the poet is using "symmetry" as an artistic metaphor, referring to the great cat's awe-inspiring beauty and terrible form. But the tiger's form and markings are also governed by symmetry in a more mathematical sense. In 1997, when delivering one of the Royal Institution's televised Christmas Lectures, I took advantage of this connection to bring a live tiger into the lecture theatre. I have never managed to create quite the same focus from the audience in any lecture since then.

Taking the term literally, the only symmetry in a tiger is an approximate bilateral symmetry, something that it shares with innumerable other living creatures, humans among them. A tiger viewed in a mirror continues to look like a tiger. But the markings on the tiger are the visible evidence of a biological process of pattern formation that is closely connected to mathematical symmetries. Nowhere is this more evident than in one of the cat's most prominent, and most geometric, features: its elegant, cylindrical tail. A series of parallel circular stripes, running round the tail, has continuous rotational symmetries, and (if extended to an infinitely long tail) discrete translational symmetries as well.

The extent to which these symmetries are real is a standard modeling issue. A real tiger's tail is furry, not a mathematical surface, and its symmetries are not exact. Nevertheless, our understanding of the cat's markings needs to explain why they have these approximate symmetries, not just explain them away by observing that they are imperfect. For the past few decades, biology has been so focused on the genetic revolution, and the molecules whose activities determine much of the form and behavior of living organisms, that it has to some extent lost sight of the organisms themselves. But now that focus is starting to change, and an old, somewhat discredited theory is being revived as a consequence.

Sixty years ago, when Francis Crick and James Watson first worked out the molecular structure of DNA (with vital input from Rosalind Franklin and Maurice Wilkins), many biologists hoped that most of the important features of a living organism could be deduced from its DNA. But, as the molecular biology revolution matured, it became clear that instead of being some fixed blueprint, DNA is more like a list of ingredients for a recipe. An awful lot depends on precisely how those ingredients are combined and cooked.

The best tool for discovering what a process does if you know its ingredients and how they interact is mathematics. So a few mavericks tried to understand the growth and form of living creatures by using mathematical techniques. Unfortunately, their ideas were overshadowed by the flood of results appearing in molecular biology, and the new ideas looked old-fashioned in comparison, so they weren't taken seriously by mainstream biologists. Recent new results suggest that this reaction was unwise.

The story starts in 1952, a year before Crick and Watson's epic discovery, when the mathematician and computing pioneer Alan Turing published a theory of animal markings in a paper with the title "The Chemical Basis of Morphogenesis." Turing is famous for his wartime code-breaking activities at Bletchley Park, the Turing test for artificial intelligence, and the undecidability of the halting problem for Turing machines. But he also worked on number theory and the markings on animals.

We are all familiar with the stripes on tigers and zebras, the spots on leopards, and the dappled patches on some breeds of cow. These patterns seldom display the exact regularity that people often expect from mathematics, but nevertheless they have a distinctly mathematical "feel." Turing modeled the formation of animal markings as a process that laid down a "pre-pattern" in the developing embryo. As the embryo grew, this pre-pattern became expressed as a pattern of protein pigments. He therefore concentrated on modeling the pre-pattern.

Turing's model has two main ingredients: reaction and diffusion. He imagined some system of chemicals, which he called morphogens.

At any given point on the part of the embryo that eventually becomes the skin—in effect, the embryo's surface—these morphogens react together to create other chemical molecules. These reactions can be modeled by ordinary differential equations. However, the skin also has a spatial structure, and that is where diffusion comes into play. The chemicals and their reaction products can also diffuse, moving across the skin in any direction. Turing wrote down partial differential equations for processes that combined these two features. We now call them reaction–diffusion equations, or Turing equations.

The most important result to emerge from Turing's model is that the reaction–diffusion process can create striking and often complex patterns. Moreover, many of these patterns are symmetric. A symmetry of a mathematical object or system is a way to transform it so that the end looks exactly the same as it started. The striped pattern on a tiger's tail, for example, has rotational and translational symmetries. It looks the same if the cylindrical tail is rotated through any angle about its axis, and it looks the same if the tail (here we assume that it is infinitely long in both directions) is translated through integer multiples of the distance between neighboring stripes at right angles to the stripes.

Turing's theory fell out of favor because it did not specify enough biological details—for example, what the morphogens actually are. Its literal interpretation also failed to predict what would happen in various experiments; for example, if the embryo developed at a different temperature. It was also realized that many different equations can produce such patterns, not just the specific ones proposed by Turing. So the occurrence of patterns like those seen in animals does not of itself confirm Turing's proposed mechanism for animal markings. Mathematically, there is a large class of equations that give the same general catalog of possible patterns. What distinguishes them is the details: which patterns occur in which circumstances. Biologists tended to see this result as an obstacle: How can you decide which equations are realistic? Mathematicians saw it as an opportunity: Let's try to find out.

Thinking along such lines, Jim Murray has developed more general versions of Turing's model and has applied them to the markings on many animals, including big cats, giraffes, and zebras. Here the iconic patterns are stripes (tiger, zebra) and spots (cheetah, leopard). Both patterns are created by wavelike structures in the chemistry. Long, parallel waves, like waves breaking on a seashore, produce stripes. A second

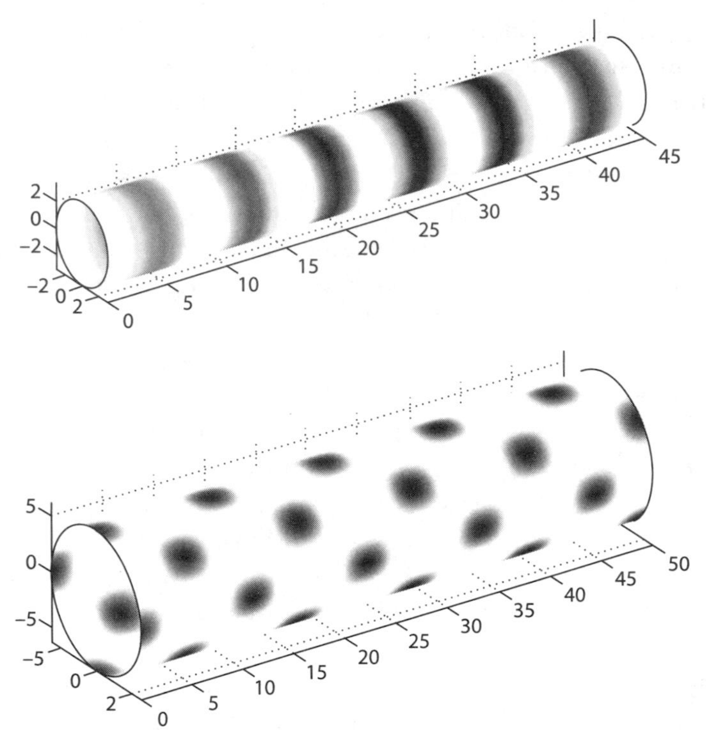

FIGURE 1. Stripes and spots in reaction–diffusion equations on a cylindrical domain. Source: http://www-rohan.sdsu.edu/~rcarrete/teaching/M-596_patt/lectures/lectures.html.

system of waves, at an angle to the first, can cause the stripes to break up into series of spots. Mathematically, stripes turn into spots when the pattern of parallel waves becomes unstable. Pursuing this pattern-making led Murray to an interesting theorem: A spotted animal can have a striped tail, but a striped animal cannot have a spotted tail.

Hans Meinhardt has studied many variants of Turing's equations, with particular emphasis on the markings on seashells. His elegant book *The Algorithmic Beauty of Seashells* studies many different kinds of chemical mechanism, showing that particular types of reactions lead to particular kinds of patterns. For example, some of the reactants inhibit the production of others, and some reactants activate the production of others. Combinations of inhibitors and activators can cause chemical oscillations, resulting in regular patterns of stripes or spots.

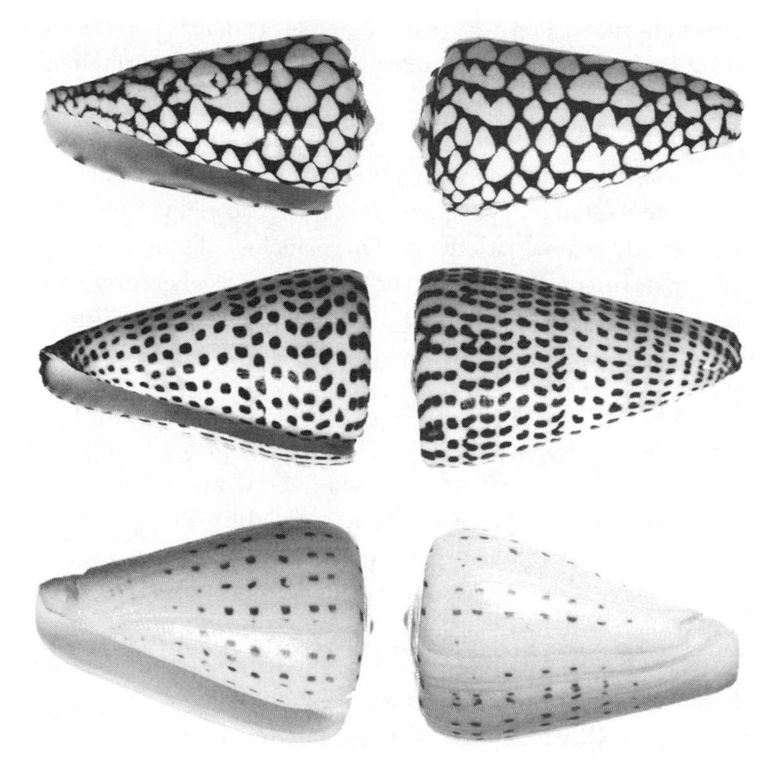

FIGURE 2. Cone shell pattern. Source: Courtesy of Shutterstock.com/Kasia

Meinhardt's theoretical patterns compare well with those found on real shells.

Stripes and spots can be obtained by solving Turing's equations numerically, and their existence can be inferred by analyzing the equations directly, using methods from bifurcation theory, which tells us what happens if a uniform pattern becomes unstable. But why are we seeing patterns at all? It is here that symmetry enters the picture. Turing himself noticed that the tendency of Turing equations to create patterns, rather than bland uniformity, can be explained using the mathematics of symmetry-breaking. In suitably mathematical regions, such as a cylinder or a plane, Turing equations are symmetric. If a solution is transformed by a symmetry operation, it remains a solution (though usually a different one). The uniform solution—corresponding to something like a lion, the same color everywhere—possesses the full set of symmetries. Stripes and spots do not, but they retain quite a lot of them.

Symmetry-breaking is an important process in pattern formation, which at first sight conflicts with a basic principle about symmetries in physical systems stated by the physicist Pierre Curie: Symmetric causes produce equally symmetric effects. This principle, if taken too literally, suggests that Turing's equations are incapable of generating patterns. In his 1952 paper, Turing includes a discussion of this point, under the heading "The Breakdown of Symmetry and Homogeneity." He writes,

> There appears superficially to be a difficulty [with the theory] . . . An embryo in its spherical blastula stage has spherical symmetry . . . A system which has spherical symmetry, and whose state is changing because of chemical reactions and diffusion, will remain spherical for ever . . . It certainly cannot result in an organism such as a horse, which is not spherically symmetrical.

However, he goes on to point out that

> There is a fallacy in this argument . . . The system may reach a state of instability in which . . . irregularities . . . tend to grow. If this happens a new and stable equilibrium is usually reached. . . . For example, if a rod is hanging from a point a little above its centre of gravity it will be in stable equilibrium. If, however, a mouse climbs up the rod, the equilibrium eventually becomes unstable and the rod starts to swing.

To drive the point home, he then provides an analogous example in a reaction–diffusion system, showing that instability of the uniform state leads to stable patterns with some—though not all—of the symmetries of the equations.

In fact, it is the reduced list of symmetries that makes the patterns visible. The tiger's stripes are visibly separated from each other because the translational symmetries are discrete. You can slide the pattern through an integer multiple of the distance between successive stripes—an integer number of wavelengths of the underlying chemical pattern.

Turing put his finger on the fundamental reason for symmetry-breaking: A fully symmetric state of the system may be unstable. Tiny perturbations can destroy the symmetry. If so, this pattern does not occur in practice. Curie's principle is not violated because the perturbations are not symmetric, but the principle is misleading because it is easy to forget about the perturbations (because they are very small). It turns out that in a symmetric system of equations, instability of the fully symmetric state usually leads to stable patterns instead.

For example, if the domain of the equations is a circle, typical broken-symmetry patterns are waves, like sine curves, with a wavelength that is the circumference of the circle divided by an integer. On a rectangular or cylindrical domain, symmetry-breaking can lead to plane waves, corresponding to stripes, and superpositions of two plane waves, corresponding to spots.

Most biologists found Turing's ideas unsatisfactory. In particular, his model did not specify what the supposed morphogens were. Biologists came to prefer a different approach to the growth and form of the embryo, known as positional information. Here an animal's body is thought of as a kind of map, and its DNA acts as an instruction book. The cells of the developing organism look at the map to find out where they are, and then at the book to find out what they should do when they are in that location. Coordinates on the map are supplied by chemical gradients: For example, a chemical might be highly concentrated near the back of the animal and gradually fade away toward the front. By "measuring" the concentration, a cell can work out where it is.

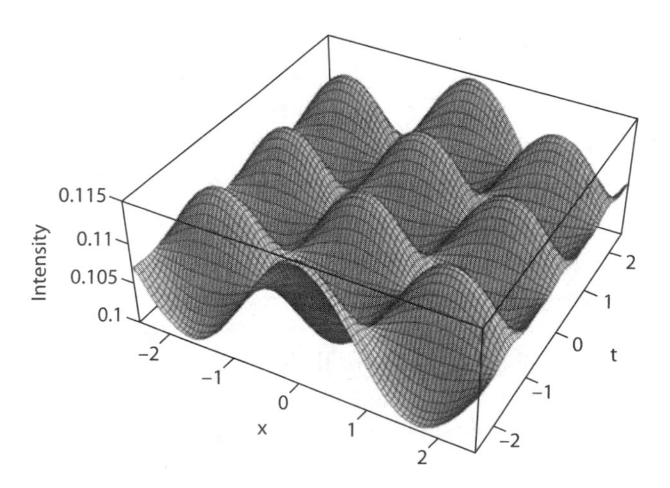

FIGURE 3. Patterns formed from one or more plane waves.

The main difference in viewpoint was that the mathematicians saw biological development as a continuous process in which the animal grew organically by following general rules controlled by specific inputs from the genes, whereas to the biologists it was more like making a model out of chemical Lego bricks by following a plan laid out in the DNA genetic instruction book.

Important evidence supporting the theory of positional information came from transplant experiments, in which tissue in a growing embryo is moved to a new location. For example, the wing bud of a chick embryo starts to develop a kind of striped pattern that later becomes the bones of the wing, and a mouse embryo starts to develop a similar pattern that eventually becomes the digits that make up its paws. The experimental results were consistent with the theory of positional information and were widely interpreted as confirming it.

Despite the apparent success of positional information, some mathematicians, engineers, physicists, and computer scientists were not convinced that a chemical gradient could provide accurate positions in a robust manner. It is now starting to look as though they were right. The experiments that seemed to confirm the positional information theory turn out to have been a little too simple to reveal some effects that are not consistent with it.

In December 2012, a team of researchers led by Rushikesh Sheth at the University of Cantabria in Spain carried out more complex transplant experiments involving a larger number of digits. They showed that a particular set of genes affects the number of digits that a growing mouse embryo develops. Strikingly, as the effect of these genes *decreases*, the mouse grows more digits than usual—like a human with six or seven fingers instead of five. This and other results are incompatible with the theory of positional information and chemical gradients, but they make complete sense in terms of Turing equations.

Other groups have discovered additional examples supporting Turing's model and have identified the specific genes and morphogens involved. In 2006, Stefanie Stick and co-workers reported experiments showing that the spacing of hair follicles (from which hairs sprout) in mice is controlled by a biochemical signaling system called WNT and proteins in the DKK family that inhibit WNT.

In 2012, a group at King's College London (Jeremy Green from the Department of Craniofacial Development at King's Dental Institute) showed that ridge patterns inside a mouse's mouth are controlled by a Turing process. The team identified the pair of morphogens working together to influence where each ridge is formed. They are FGF (fibroblast growth factor) and Shh (Sonic Hedgehog, so called because fruit flies that lack the corresponding gene look spiky).

These results are likely to be followed by many others that provide genetic underpinning for Turing's model or more sophisticated variants. They show that genetic information alone cannot explain the structural changes that occur in a developing embryo. More flexible mathematical processes are also important. On the other hand, mathematical models become genuinely useful only when they are combined with detailed and specific information about which genes are active in a given process. Neither molecular biology nor mathematics alone can explain the markings on animals, or how a developing organism changes as it grows. It will take both, working together, to frame the fearful symmetry of the tyger.

References

Andrew D. Economou, Atsushi Ohazama, Thantrira Porntaveetus, Paul T. Sharpe, Shigeru Kondo, M. Albert Basson, Amel Gritli-Linde, Martyn T. Cobourne, and Jeremy B. A. Green, "Periodic stripe formation by a Turing mechanism operating at growth zones in the mammalian palate." *Nature Genetics* **44** (2012) 348–351; doi: 10.1038/ng.1090.

Rushikesh Sheth, Luciano Marcon, M. Félix Bastida, Marisa Junco, Laura Quintana, Randall Dahn, Marie Kmita, James Sharpe, and Maria A. Ros, "*Hox* genes regulate digit patterning by controlling the wavelength of a Turing-type mechanism." *Science* **338** (2012) 6113, 1476–1480; doi: 10.1126/science.1226804; available at http://www.sciencemag.org/content/338/6113/1476.full?sid=dc2a12e8-c1c7-4f2a-bc4e-97962850caa3.

Alan Turing, "The chemical basis of morphogenesis." *Phil. Trans. R. Soc. London B* **237** (1952) 37–72.

E pluribus unum:
From Complexity, Universality

TERENCE TAO

Nature is a mutable cloud, which is always and never the same.
—Ralph Waldo Emerson, "History" (1841)

Modern mathematics is a powerful tool to model any number of real-world situations, whether they be natural—the motion of celestial bodies, for example, or the physical and chemical properties of a material—or man-made: for example, the stock market or the voting preferences of an electorate.[1] In principle, mathematical models can be used to study even extremely complicated systems with many interacting components. However, in practice, only simple systems (ones that involve only two or three interacting agents) can be solved precisely. For instance, the mathematical derivation of the spectral lines of hydrogen, with its single electron orbiting the nucleus, can be given in an undergraduate physics class; but even with the most powerful computers, a mathematical derivation of the spectral lines of sodium, with 11 electrons interacting with each other and with the nucleus, is out of reach. (The *three-body problem,* which asks to predict the motion of three masses with respect to Newton's law of gravitation, is famously known as the only problem to have ever given Newton headaches. Unlike the two-body problem, which has a simple mathematical solution, the three-body problem is believed not to have any simple mathematical expression for its solution and can only be solved approximately, via numerical algorithms.) The inability to perform feasible computations on a system with many interacting components is known as the *curse of dimensionality.*

Despite this curse, a remarkable phenomenon often occurs once the number of components becomes large enough: that is, the aggregate properties of the complex system can mysteriously become predictable again, governed by simple laws of nature. Even more surprisingly, these macroscopic laws for the overall system are often largely *independent* of their microscopic counterparts that govern the individual components of that system. One could replace the microscopic components by completely different types of objects and obtain the same governing law at the macroscopic level. When this phenomenon occurs, we say that the macroscopic law is *universal.* The universality phenomenon has been observed both empirically and mathematically in many different contexts, several of which I discuss below. In some cases, the phenomenon is well understood, but in many situations, the underlying source of universality is mysterious and remains an active area of mathematical research.

The U.S. presidential election of November 4, 2008, was a massively complicated affair. More than 100 million voters from 50 states cast their ballots; each voter's decision was influenced in countless ways by campaign rhetoric, media coverage, rumors, personal impressions of the candidates, political discussions with friends and colleagues, or all of these things. There were millions of "swing" voters who were not firmly supporting either of the two major candidates; their final decisions would be unpredictable and perhaps even random in some cases. The same uncertainty existed at the state level: Although many states were considered safe for one candidate or the other, at least a dozen were considered "in play" and could have gone either way.

In such a situation, it would seem impossible to forecast accurately the election outcome. Sure, there were electoral polls—hundreds of them—but each poll surveyed only a few hundred or a few thousand likely voters, which is only a tiny fraction of the entire population. And the polls often fluctuated wildly and disagreed with each other; not all polls were equally reliable or unbiased, and no two polling organizations used exactly the same methodology.

Nevertheless, well before election night was over, the polls had predicted the outcome of the presidential election (and most other elections taking place that night) quite accurately. Perhaps most spectacular were the predictions of statistician Nate Silver, who used a weighted analysis of all existing polls to predict correctly the outcome of the presidential election in 49 out of 50 states, as well as in all of the 35

U.S. Senate races. (The lone exception was the presidential election in Indiana, which Silver called narrowly for McCain but which eventually favored Obama by just 0.9 percent.)

The accuracy of polling can be explained by a mathematical law known as the *law of large numbers*. Thanks to this law, we know that once the size of a random poll is large enough, the probable outcomes of that poll will converge to the actual percentage of voters who would vote for a given candidate, up to a certain accuracy, known as the *margin of error*. For instance, in a random poll of 1,000 voters, the margin of error is about 3 percent.

A remarkable feature of the law of large numbers is that it is *universal*. Does the election involve 100,000 voters or 100 million voters? It doesn't matter: The margin of error for the poll will remain 3 percent. Is it a state that favors McCain to Obama 55 percent to 45 percent, or Obama to McCain 60 percent to 40 percent? Is the state a homogeneous bloc of, say, affluent white urban voters, or is the state instead a mix of voters of all incomes, races, and backgrounds? Again, it doesn't matter: The margin of error for the poll will still be 3 percent. The only factor that makes a significant difference is the size of the poll; the larger the poll, the smaller the margin of error. The immense complexity of 100 million voters trying to decide between presidential candidates collapses to just a handful of numbers.

The law of large numbers is one of the simplest and best understood of the universal laws in mathematics and nature, but it is by no means the only one. Over the decades, many such universal laws have been found to govern the behavior of wide classes of complex systems, regardless of the components of a system or how they interact with each other.

In the case of the law of large numbers, the mathematical underpinnings of the universality phenomenon are well understood and are taught routinely in undergraduate courses on probability and statistics. However, for many other universal laws, our mathematical understanding is less complete. The question of why universal laws emerge so often in complex systems is a highly active direction of research in mathematics. In most cases, we are far from a satisfactory answer to this question, but as I discuss below, we have made some encouraging progress.

After the law of large numbers, perhaps the next most fundamental example of a universal law is the *central limit theorem*. Roughly speaking, this theorem asserts that if one takes a statistic that is a combination of many independent and randomly fluctuating components, with no one component having a decisive influence on the whole, then that statistic will be approximately distributed according to a law called the *normal distribution* (or *Gaussian distribution)* and more popularly known as the *bell curve*. The law is universal because it holds regardless of exactly how the individual components fluctuate or how many components there are (although the accuracy of the law improves when the number of components increases). It can be seen in a staggeringly diverse range of statistics, from the incidence rate of accidents; to the variation of height, weight, or other vital statistics among a species; to the financial gains or losses caused by chance; to the velocities of the component particles of a physical system. The size, width, location, and even the units of measurement of the distribution vary from statistic to statistic, but the bell curve shape can be discerned in all cases. This convergence arises not because of any low-level, or microscopic, connection among such diverse phenomena as car crashes, human height, trading profits, or stellar velocities, but because in all these cases the high-level, or macroscopic, structure is the same: namely, a compound statistic formed from a combination of the small influences of many independent factors. That the macroscopic behavior of a large, complex system can be almost totally independent of its microscopic structure is the essence of universality.

The universal nature of the central limit theorem is tremendously useful in many industries, allowing them to manage what would otherwise be an intractably complex and chaotic system. With this theorem, insurers can manage the risk of, say, their car insurance policies without having to know all the complicated details of how car crashes occur; astronomers can measure the size and location of distant galaxies without having to solve the complicated equations of celestial mechanics; electrical engineers can predict the effect of noise and interference on electronic communications without having to know exactly how this noise is generated; and so forth. The central limit theorem, though, is not completely universal; there are important cases when the theorem does not apply, giving statistics with a distribution quite different from the bell curve. (I will return to this point later.)

There are distant cousins of the central limit theorem that are universal laws for slightly different types of statistics. One example, *Benford's law*, is a universal law for the first few digits of a statistic of large magnitude, such as the population of a country or the size of an account; it gives a number of counterintuitive predictions: for instance, that any given statistic occurring in nature is more than six times as likely to start with the digit 1 than with the digit 9. Among other things, this law (which can be explained by combining the central limit theorem with the mathematical theory of logarithms) has been used to detect accounting fraud because numbers that are made up, as opposed to those that arise naturally, often do not obey Benford's law (Figure 1).

In a similar vein, *Zipf's law* is a universal law that governs the largest statistics in a given category, such as the largest country populations in the world or the most frequent words in the English language. It asserts that the size of a statistic is usually inversely proportional to its ranking; thus, for instance, the tenth largest statistic should be about half the size of the fifth largest statistic. (The law tends not to work so well for the top two or three statistics, but becomes more accurate after that.) Unlike the central limit theorem and Benford's law, which are fairly well understood mathematically, Zipf's law is primarily an empirical law; it is observed in practice, but mathematicians do not have a fully satisfactory and convincing explanation for how the law comes about and why it is universal.

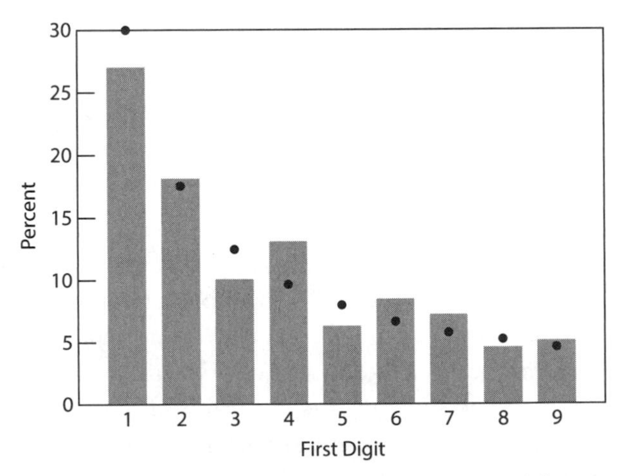

FIGURE 1. Histogram of the first digits of the populations of the 237 countries of the world in 2010. The black dots indicate the Benford's law prediction. Source: Wikipedia, http://en.wikipedia.org/wiki/File:Benfords_law_illustrated_by_world%27s_countries_population.png; used here under The Creative Commons Attribution-ShareAlike license.

So far, I have discussed universal laws for individual statistics: complex numerical quantities that arise as the combination of many smaller and independent factors. But universal laws have also been found for more complicated objects than mere numerical statistics. Take, for example, the laws governing the complicated shapes and structures that arise from *phase transitions* in physics and chemistry. As we learn in high school science classes, matter comes in various states, including the three classic states of solid, liquid, and gas, but also a number of exotic states such as plasmas or superfluids. Ferromagnetic materials, such as iron, also have magnetized and nonmagnetized states; other materials become electrical conductors at some temperatures and insulators at others. What state a given material is in depends on a number of factors, most notably the temperature and, in some cases, the pressure. (For some materials, the level of impurities is also relevant.) For a fixed value of the pressure, most materials tend to be in one state for one range of temperatures and in another state for another range. But when the material is at or very close to the temperature dividing these two ranges, interesting phase transitions occur. The material, which is not fully in one state or the other, tends to split into beautifully fractal shapes known as clusters, each of which embodies one or the other of the two states.

There are countless materials in existence, each with a different set of key parameters (such as the boiling point at a given pressure). There are also a large number of mathematical models·that physicists and chemists use to model these materials and their phase transitions, in which individual atoms or molecules are assumed to be connected to some of their neighbors by a random number of bonds, assigned according to some probabilistic rule. At the microscopic level, these models can look quite different from each other. For instance, Figures 2 and 3 display the small-scale structure of two typical models. Figure 2 shows a site percolation model on a hexagonal lattice in which each hexagon (or site) is an abstraction of an atom or molecule randomly placed in one of two states; the clusters are the connected regions of a single color. Figure 3 shows a bond percolation model on a square lattice in which the edges of the lattice are abstractions of molecular bonds that each have some probability of being activated; the clusters are the connected regions given by the active bonds.

If one zooms out to look at the large-scale structure of clusters while at or near the critical value of parameters (such as temperature),

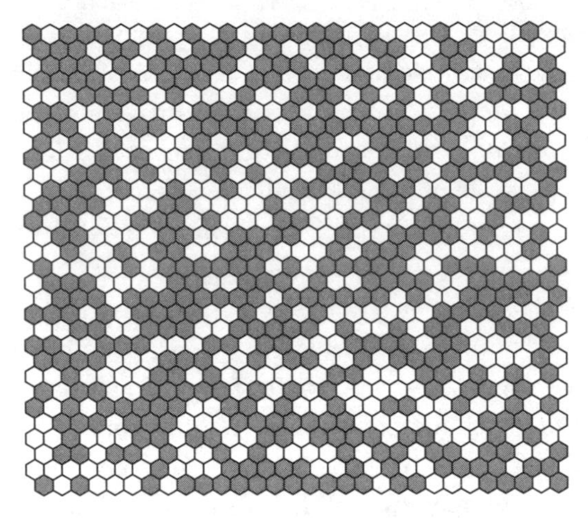

FIGURE 2. Site percolation model on a hexagonal lattice at the critical threshold. Source: Michael Kozdron, http://stat.math.uregina.ca/~kozdron/Simulations/Percolation/percolation.html; used here with permission from Michael Kozdron.

the differences in microscopic structure fade away, and one begins to see a number of universal laws emerging. Although the clusters have a random size and shape, they almost always have a fractal structure; thus, if one zooms in on any portion of the cluster, the resulting image more or less resembles the cluster as a whole. Basic statistics, such as the number of clusters, the average size of the clusters, or the frequency with which a cluster connects two given regions of space, appear to obey some specific universal laws, known as *power laws* (which are somewhat similar, though not quite the same, as Zipf's law). These laws arise in almost every mathematical model that has been put forward to explain (continuous) phase transitions and have been observed many times in nature. As with other universal laws, the precise microscopic structure of the model or the material may affect some basic parameters, such as the phase transition temperature, but the underlying structure of the law is the same across all models and materials.

In contrast to more classical universal laws such as the central limit theorem, our understanding of the universal laws of phase transition is incomplete. Physicists have put forth some compelling heuristic arguments that explain or support many of these laws (based on a powerful,

FIGURE 3. Bond percolation model on a square lattice at the critical threshold. Note the presence of both very small clusters and extremely large clusters. Source: Wikipedia, http://en.wikipedia.org/wiki/File:Bond_percolation_p_51.png; used here under the Creative Commons Attribution-ShareAlike license.

but not fully rigorous, tool known as the *renormalization group method),* but a completely rigorous proof of these laws has not yet been obtained in all cases. This field is a very active area of research; for instance, in August 2010, the Fields medal, one of the most prestigious prizes in mathematics, was awarded to Stanislav Smirnov for his breakthroughs in rigorously establishing the validity of these universal laws for some key models, such as percolation models on a triangular lattice.

As we near the end of our tour of universal laws, I want to consider an example of this phenomenon that is closer to my own area of research. Here, the object of study is not a single numerical statistic (as in the case of the central limit theorem) or a shape (as with phase transitions), but a discrete spectrum: a sequence of points (or numbers, or frequencies, or energy levels) spread along a line.

Perhaps the most familiar example of a discrete spectrum is the radio frequencies emitted by local radio stations; this set is a sequence of frequencies in the radio portion of the electromagnetic spectrum, which one can access by turning a radio dial. These frequencies are not evenly spaced, but usually some effort is made to keep any two station frequencies separated from each other to reduce interference.

Another familiar example of a discrete spectrum is the spectral lines of an atomic element, which come from the frequencies that the electrons in the atomic shells can absorb and emit, according to the laws of quantum mechanics. When these frequencies lie in the visible portion of the electromagnetic spectrum, they give individual elements their distinctive colors, from the blue light of argon gas (which, confusingly, is often used in neon lamps because pure neon emits orange-red light) to the yellow light of sodium. For simple elements, such as hydrogen, the equations of quantum mechanics can be solved relatively easily, and the spectral lines follow a regular pattern; however, for heavier elements, the spectral lines become quite complicated and not easy to work out just from first principles.

An analogous, but less familiar, example of spectra comes from the scattering of neutrons off of atomic nuclei, such as the uranium-238 nucleus. The electromagnetic and nuclear forces of a nucleus, when combined with the laws of quantum mechanics, predict that a neutron will pass through a nucleus virtually unimpeded for some energies but will bounce off that nucleus at other energies, known as scattering resonances. The internal structures of such large nuclei are so complex that it has not been possible to compute these resonances either theoretically or numerically, leaving experimental data as the only option.

These resonances have an interesting distribution; they are not independent of each other but instead seem to obey a precise repulsion law that makes it unlikely that two adjacent resonances are too close to each other—somewhat analogous to how radio station frequencies tend to avoid being too close together, except that the former phenomenon arises from the laws of nature rather than from government regulation of the spectrum. In the 1950s, the renowned physicist and Nobel laureate Eugene Wigner investigated these resonance statistics and proposed a remarkable mathematical model to explain them, an example of what we now call a *random matrix model.* The precise mathematical details of these models are too technical to describe here, but in general, one can view such models as a large collection of masses, all connected to each other by springs of various randomly selected strengths. Such a mechanical system will oscillate (or resonate) at a certain set of frequencies; and the Wigner hypothesis asserts that the resonances of a large atomic nucleus should resemble that of the resonances of a random matrix model. In particular, they should experience the same repulsion phenomenon. Because it is possible to rigorously prove repulsion of the frequencies of a random matrix model, a heuristic explanation can be given for the same phenomenon that is experimentally observed for nuclei.

Of course, an atomic nucleus does not actually resemble a large system of masses and springs (among other things, it is governed by the laws of quantum mechanics rather than of classical mechanics). Instead, as we have since discovered, Wigner's hypothesis is a manifestation of a universal law that governs many types of spectral lines, including those that ostensibly have little in common with atomic nuclei or random matrix models. For instance, the same spacing distribution was famously found in the waiting times between buses arriving at a bus stop in Cuernavaca, Mexico (without, as yet, a compelling explanation for why this distribution emerges in this case).

Perhaps the most unexpected demonstration of the universality of these laws came from the wholly unrelated area of *number theory,* and in particular the distribution of the prime numbers 2, 3, 5, 7, 11, and so on—the natural numbers greater than 1that cannot be factored into smaller natural numbers. The prime numbers are distributed in an irregular fashion through the integers; but if one performs a spectral analysis of this distribution, one can discern certain long-term oscillations in the distribution (sometimes known as the music of the primes), the frequencies of which are described by a sequence of complex numbers known as the (nontrivial) zeroes of the Riemann zeta function, first studied by Bernhard Riemann in 1859. (For this discussion, it is not important to know exactly what the Riemann zeta function is.) In principle, these numbers tell us everything we would wish to know about the primes. One of the most famous and important problems in number theory is the *Riemann hypothesis,* which asserts that these numbers all lie on a single line in the complex plane. It has many consequences in number theory and, in particular, gives many important consequences about the prime numbers. However, even the powerful Riemann hypothesis does not settle everything on this subject, in part because it does not directly say much about how the zeroes are distributed on this line. But there is extremely strong numerical evidence that these zeroes obey the same precise law that is observed in neutron scattering and in other systems; in particular, the zeroes seem to "repel" each other in a manner that matches the predictions of random matrix theory with uncanny accuracy. The formal description of this law is known as the Gaussian unitary ensemble (GUE) hypothesis. (The GUE hypothesis is a fundamental example of a random matrix model.) Like the Riemann hypothesis, it is currently unproven, but it has powerful consequences for the distribution of the prime numbers.

The discovery of the GUE hypothesis, connecting the music of the primes and the energy levels of nuclei, occurred at the Institute for Advanced Study in 1972, and the story is legendary in mathematical circles. It concerns a chance meeting between the mathematician Hugh Montgomery, who had been working on the distribution of zeroes of the zeta function (and more specifically, on a certain statistic relating to that distribution known as the pair correlation function), and the renowned physicist Freeman Dyson. In his book *Stalking the Riemann Hypothesis,* mathematician and computer scientist Dan Rockmore describes that meeting:

> As Dyson recalls it, he and Montgomery had crossed paths from time to time at the Institute nursery when picking up and dropping off their children. Nevertheless, they had not been formally introduced. In spite of Dyson's fame, Montgomery hadn't seen any purpose in meeting him. "What will we talk about?" is what Montgomery purportedly said when brought to tea. Nevertheless, Montgomery relented and upon being introduced, the amiable physicist asked the young number theorist about his work. Montgomery began to explain his recent results on the pair correlation, and Dyson stopped him short—"Did you get this?" he asked, writing down a particular mathematical formula. Montgomery almost fell over in surprise: Dyson had written down the sinc-infused pair correlation function. . . . Whereas Montgomery had traveled a number theorist's road to a "prime picture" of the pair correlation, Dyson had arrived at this formula through the study of these energy levels in the mathematics of matrices.[2]

The chance discovery by Montgomery and Dyson that the same universal law that governs random matrices and atomic spectra also applies to the zeta function was given substantial numerical support by the computational work of Andrew Odlyzko beginning in the 1980s (Figure 4). But this discovery does not mean that the primes are somehow nuclear-powered or that atomic physics is somehow driven by the prime numbers; instead, it is evidence that a single law for spectra is so universal that it is the natural end product of any number of different processes, whether from nuclear physics, random matrix models, or number theory.

The precise mechanism underlying this law has not yet been fully unearthed; in particular, we still do not have a compelling explanation, let alone a rigorous proof, of why the zeroes of the zeta function are subject to the GUE hypothesis. However, there is now a substantial body of rigorous work (including some of my own work, and including some substantial breakthroughs in just the past few years) that gives

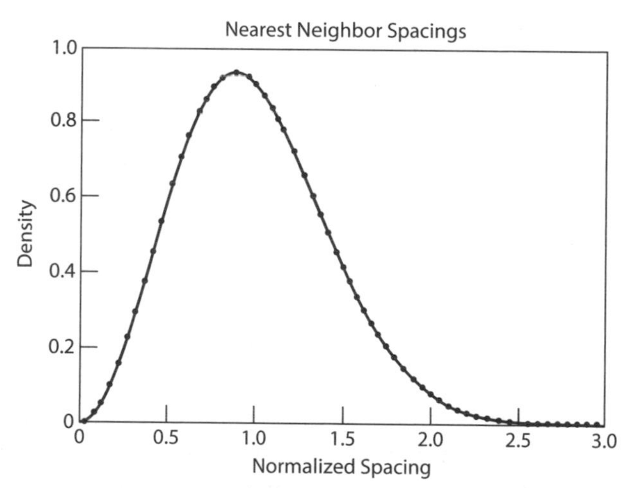

FIGURE 4. Spacing distribution for a billion zeroes of the Riemann zeta function, with the corresponding prediction from random matrix theory. Source: Andrew M. Odlyzko, "The 10^{22}nd Zero of the Riemann Zeta Function," in *Dynamical, Spectral, and Arithmetic Zeta Functions*, Contemporary Mathematics Series, no. 290, ed. Machiel van Frankenhuysen and Michel L. Lapidus (American Mathematical Society, 2001), 139–144, http://www.dtc.umn.edu/~odlyzko/doc/zeta.10to22.pdf; used here with permission from Andrew Odlyzko.

support to the universality of this hypothesis by showing that a wide variety of random matrix models (not just the most famous model of the GUE) are all governed by essentially the same law for their spacings. At present, these demonstrations of universality have not extended to the number theoretic or physical settings, but they do give indirect support to the law being applicable in those cases.

The arguments used in this recent work are too technical to give here, but I do mention one of the key ideas, which my colleague Van Vu and I borrowed from an old proof of the central limit theorem by Jarl Lindeberg from 1922. In terms of the mechanical analogy of a system of masses and springs (mentioned above), the key strategy was to replace just one of the springs by another, randomly selected spring and to show that the distribution of the frequencies of this system did not change significantly when doing so. Applying this replacement operation to each spring in turn, one can eventually replace a given random matrix model with a completely different model while keeping the distribution mostly unchanged—which can be used to show that large classes of random matrix models have essentially the same distribution.

This field is a very active area of research; for instance, simultaneously with Van Vu's and my work from last year, László Erdös, Benjamin Schlein, and Horng-Tzer Yau also gave a number of other demonstrations of universality for random matrix models, based on ideas from mathematical physics. The field is moving quickly, and in a few years we may have many more insights into the nature of this mysterious universal law.

There are many other universal laws of mathematics and nature; the examples I have given are only a small fraction of those that have been discovered over the years, from such diverse subjects as dynamical systems and quantum field theory. For instance, many of the macroscopic laws of physics, such as the laws of thermodynamics or the equations of fluid motion, are quite universal in nature, making the microscopic structure of the material or fluid being studied almost irrelevant, other than via some key parameters, such as viscosity, compressibility, or entropy.

However, the principle of universality does have definite limitations. Take, for instance, the central limit theorem, which gives a bell curve distribution to any quantity that arises from a combination of many small and independent factors. This theorem can fail when the required hypotheses are not met. The distribution of, say, the heights of all human adults (male and female) does not obey a bell curve distribution because one single factor—gender—has so large an effect on height that it is not averaged out by all the other environmental and genetic factors that influence this statistic.

Another important way in which the central limit fails is when the individual factors that make up a quantity do not fluctuate independently of each other but are instead correlated, so that they tend to rise or fall in unison. In such cases, "fat tails" (also known colloquially as "black swans") can develop, in which the quantity moves much further from its average value than the central limit theorem would predict. This phenomenon is particularly important in financial modeling, especially when dealing with complex financial instruments such as the collateralized debt obligations that were formed by aggregating mortgages. As long as the mortgages behaved independently of each other, the central limit theorem could be used to model the risk of these instruments. However, in the recent financial crisis (a textbook example of a black swan), this independence hypothesis broke down spectacularly, leading to significant financial losses for many holders of these obligations (and for their insurers). A mathematical model is only as strong as the assumptions behind it.

A third way in which a universal law can break down is if the system does not have enough degrees of freedom for the law to take effect. For instance, cosmologists can use universal laws of fluid mechanics to describe the motion of entire galaxies, but the motion of a single satellite under the influence of just three gravitational bodies can be far more complicated (quite literally, rocket science).

Another instance where the universal laws of fluid mechanics break down is at the *mesoscopic* scale: that is, larger than the microscopic scale of individual molecules but smaller than the macroscopic scales for which universality applies. An important example of a mesoscopic fluid is the blood flowing through blood vessels; the blood cells that make up this liquid are so large that they cannot be treated merely as an ensemble of microscopic molecules but rather as mesoscopic agents with complex behavior. Other examples of materials with interesting mesoscopic behavior include colloidal fluids (such as mud), certain types of nanomaterials, and quantum dots; it is a continuing challenge to mathematically model such materials properly.

There are also many macroscopic situations in which no universal law is known to exist, particularly in cases where the system contains human agents. The stock market is a good example: Despite extremely intensive efforts, no satisfactory universal laws to describe the movement of stock prices have been discovered. (The central limit theorem, for instance, does not seem to be a good model, as discussed earlier.) One reason for this shortcoming is that any regularity discovered in the market is likely to be exploited by arbitrageurs until it disappears. For similar reasons, finding universal laws for macroeconomics appears to be a moving target; according to *Goodhart' s law*, if an observed statistical regularity in economic data is exploited for policy purposes, it tends to collapse. (Ironically, Goodhart's law itself is arguably an example of a universal law.)

Even when universal laws do exist, it still may be practically impossible to use them to make predictions. For instance, we have universal laws for the motion of fluids, such as the Navier–Stokes equations, and these are used all the time in such tasks as weather prediction. But these equations are so complex and *unstable* that even with the most powerful computers, we are still unable to accurately predict the weather more than a week or two into the future. (By unstable, I mean that even small errors in one's measurement data, or in one's numerical computations, can lead to large fluctuations in the predicted solution of the equations.)

Hence, between the vast, macroscopic systems for which universal laws hold sway and the simple systems that can be analyzed using the fundamental laws of nature, there is a substantial middle ground of systems that are too complex for fundamental analysis but too simple to be universal—plenty of room, in short, for all the complexities of life as we know it.

Notes

1. This essay benefited from the feedback of many readers of my blog. They commented on a draft version that (together with additional figures and links) can be read at http://terrytao.wordpress.com/2010/09/14/, A second draft of a non-technical article on universality.

2. Dan Rockmore, *Stalking the Riemann Hypothesis: The Quest to Find the Hidden Law of Prime Numbers* (New York: Pantheon Books, 2005).

Degrees of Separation

Gregory Goth

The idea of six degrees of separation—that is, that every person in the world is no more than six people away from every other person on earth—has fascinated social scientists and laymen alike ever since Hungarian writer Frigyes Karinthy introduced the concept in 1929. (The story has since been reprinted in 2006 in Princeton University Press's *The Structure and Dynamics of Networks* by Mark Newman, Albert-László Barabási, and Duncan J. Watts, translated by Adam Makkai and edited by Enikö Jankó.)

For the greater public, the cultural touchstone of the theory was the 1990 play entitled *Six Degrees of Separation* by John Guare. Although the drama was not an exploration of the phenomenon by any means, it spawned countless versions of parlor games. For scientists, however, the wellspring of the six degrees phenomenon, also called the small-world problem, was a 1967 study undertaken by social psychologist Stanley Milgram, in which a selected group of volunteers in the midwestern United States was instructed to forward messages to a target person in Boston. Milgram's results, published in *Psychology Today* in 1967, were that the messages were delivered by "chains" that comprised anywhere between 2 and 10 intermediaries; the mean was 5 (Figure 1).

In the ensuing years, the problem has become a perennial favorite among researchers of many disciplines, from computer scientists exploring probabilistic algorithms for best use of network resources to epidemiologists exploring the interplay of infectious diseases and network theory.

Most recently, the vast architectural resources of Facebook and Twitter have supplied researchers with something they never possessed before—the capability to look at the small-world problem from both the traditional algorithmic approach, which explores the probabilities of how each person (or network node) in a chain seeks out the next messenger using only the limited local knowledge they possess, and the new topological approach, which can examine the entire structure of a network as it also observes the progression of the algorithmic chains.

"It's amazing how far we've come," says Duncan Watts, a founding partner at Microsoft Research New York City, who was until recently a senior researcher at Yahoo! Watts is one of the world's leading authorities on the small-world problem, dating to the publication of "Collective Dynamics of 'Small-World' Networks," coauthored with Steven Strogatz, in *Nature* in 1998. At that time, Watts says, the largest available network, actors listed in the Internet Movie Database, contained about 225,000 edge nodes (individual actors). A recent study by researchers from Facebook and the University of Milan, however, looked at 721 million Facebook users, who had 69 billion unique friendships among them, and revealed an average of 3.74 intermediaries between a source and target user, suggesting an even smaller world than Milgram's original study showed.

"In fact, the whole motivation of the thing I did with Strogatz was precisely that you couldn't do the exercise Facebook just did," Watts says. "Now the empirical exercise is possible. That's a remarkable change."

A Similarity of Results

One must consider the large variety of online communities and compare the small-world experiments performed on them to Milgram's method—sending a message via terrestrial delivery routes—in order to fully appreciate the similarity of results across the board. Whereas the Facebook experiment yielded approximately four degrees of separation, work by distinguished scientist Eric Horvitz of Microsoft Research and Stanford University assistant professor Jure Leskovec, on connections between users of the Microsoft Instant Messaging network, yielded an average 6.6 degrees of separation between any two users. In their 2009 paper "Social Search in 'Small-World' Experiments" examining the algorithmic approach, Watts, Sharad Goel, and Roby Muhamad discovered that roughly half of all chains can be completed in six or seven steps, "thus supporting the 'six degrees of separation' assertion," they wrote, "but on the other hand, estimates of the mean are much longer, suggesting that for at least some of the population, the world is not 'small' in the algorithmic sense."

Discovering the reason why "the world is not 'small' in the algorithmic sense" presents a wide swath of fertile ground for those researchers, including Watts and Leskovec, who are still plumbing the many vectors of network navigation.

One ironic, or counterintuitive, factor in examining the small-world problem as online communities grow ever larger is that the experiments' attrition rates are also vastly greater than in the past. For instance, Watts says that only 12% of those who signed up for a joint small-world experiment at Yahoo! and Facebook completed their chains, compared with 75% of those who participated in Milgram's experiment and the 35% who completed chains in a 2001–2002 experiment run by Watts.

However, Watts says the data they have should allow them still to answer the questions they care about most, which are about exploring the efficiency of intermediary connections selected.

"We know how far you are from the target, Facebook knows how far your friends are from the target, and we know who you picked, so we can establish whether you made the right choice," Watts says. "So we can get most of the science out of it, it's just a little bummer that the attrition was so bad."

The logic behind finding the most efficient paths may produce payoffs unforeseen for both theoretical modeling and production networks such as search engine optimization. Finding the best ways to determine those paths, though, will necessitate a leap from the known models of small-world networks to a better understanding of the intermediary steps between any two endpoints of a chain.

Leskovec says that, given constants from graph theory, the diameter of any given network will grow logarithmically with its size; that is, the difference between five and six degrees of separation mandates a graph an order of magnitude larger or denser. Jon Kleinberg, Tisch University professor in the department of computer science at Cornell University, whose "The Small-World Phenomenon: An Algorithmic Perspective" is regarded as one of the problem's seminal modeling documents, says this basic property is precisely what makes the small-world theory so appealing while also presenting the research community the greatest challenge inherent in it.

"It's something that still feels counterintuitive when you first encounter it," Kleinberg says. "It makes sense in the end: I know 1,000 people, and my friend knows 1,000 people—and you don't have to multiply 1,000 by itself too many times for it to make sense."

However, this logarithmic progression also precludes the ability to examine or design intermediate levels of scale, Kleinberg says. "We thought the right definition of distance was going to be 'Here I am, and how many steps do I have to go to get to you?' but that turns out not to be. We need some other measure and I think that remains an interesting open question that people are actively looking at: Is there some kind of smoother scale here? Who are the 10,000 people closest to me? The 100,000?

"We need a much more subtle way to do that, and it is going to require some sophisticated mathematical ideas and sophisticated combinational ideas—what is the right definition of distance when you're looking at social networks? It's not just how many steps I have to go. That's an important question in everyday life and when you're designing some online system."

Mozart Meets the Terminator

Recent research is beginning to use the short-path principles of social search in the online systems discussed by Kleinberg. In "Degrees of Separation in Social Networks," presented at the Fourth International Symposium on Combinatorial Search 2011, researchers from Shiraz University, Carnegie Mellon University, and the University of Alberta designed a search algorithm, tested on Twitter, intended for uses beyond social search.

For example, they reported in Voice over Internet Protocol (VoIP) networks, when a user calls another user in the network, he or she is first connected to a VoIP carrier, a main node in the network. The VoIP carrier connects the call to the destination either directly or, more commonly, through another VoIP carrier.

"The length of the path from the caller to the receiver is important since it affects both the quality and price of the call," the researchers noted. "The algorithms that are developed in this paper can be used to find a short path (fewest carriers) between the initial (sender) and the goal (receiver) nodes in the network."

These algorithms, such as greedy algorithms enhanced by geographic heuristics, or probabalistic bidirectional methods, have the potential to cut some of the overhead, and cost, of network search sessions, such as the sample VoIP session, the authors believe.

Leskovec's most recent work based on small-world algorithms explores the paths that humans take in connecting concepts that, on the surface, seem rather disparate, such as Wolfgang Amadeus Mozart and the *Terminator* character from the science-fiction films starring Arnold Schwarzenegger.

"As a human, I sort of know how the knowledge fits together," Leskovec says. "If I want to go from Mozart to *Terminator* and I know Mozart was from Austria and Schwarzenegger was from Austria, maybe I can go through the Austrian connection. A computer that is truly decentralized has no clue; it has no conception that getting to Schwarzenegger is good enough."

Interestingly enough, Leskovec says, computers fared better than humans on average on solving such search chains, but humans also were less likely to get totally lost and were capable of forming backup plans, which the Web-crawling agents could not do. Effectively, he says, the payoff of such research is "understanding how humans do this, what kind of cues are we using, and how to make the cues more efficient or help us recognize them, to help us understand where we are, right now, in this global network."

Further Reading

Backstrom, L., Boldi, P., Rosa, M., Ugander, J., and Vigna, S. "Four degrees of separation," http://arxiv.org/abs/1111.4570, Jan. 6, 2012.

Bakhshandeh, R., Samadi, M., Azimifar, Z., and Schaeffer, J. "Degrees of separation in social networks," *Proceedings of the Fourth International Symposium on Combinatorial Search,* Barcelona, Spain, July 15–16, 2011.

Goel, S., Muhamad, R., and Watts, D. "Social search in 'small-world' experiments," 18th International World Wide Web Conference, Madrid, Spain, April 20–24, 2009.

Kleinberg, J. "The small-world phenomenon: An algorithmic perspective," 32nd ACM Symposium on Theory of Computing, Portland, OR, May 21–23, 2000.

West, R., and Leskovec, J. "Human wayfinding in information networks," 22nd International World Wide Web Conference, Lyon, France, April 16–20, 2012.

Randomness

Charles Seife

Our very brains revolt at the idea of randomness. We have evolved as a species to become exquisite pattern-finders; long before the advent of science, we figured out that a salmon-colored sky heralds a dangerous storm or that a baby's flushed face likely means a difficult night ahead. Our minds automatically try to place data in a framework that allows us to make sense of our observations and use them to understand and predict events.

Randomness is so difficult to grasp because it works against our pattern-finding instincts. It tells us that sometimes there is no pattern to be found. As a result, randomness is a fundamental limit to our intuition; it says that there are processes we can't predict fully. It's a concept that we have a hard time accepting, even though it's an essential part of the way the cosmos works. Without an understanding of randomness, we are stuck in a perfectly predictable universe that simply doesn't exist outside our heads.

I would argue that only once we understand three dicta—three laws of randomness—can we break out of our primitive insistence on predictability and appreciate the universe for what it is, rather than what we want it to be.

The First Law of Randomness: There Is Such a Thing as Randomness

We use all kinds of mechanisms to avoid confronting randomness. We talk about karma, in a cosmic equalization that ties seemingly unconnected events together. We believe in runs of luck, both good and ill, and that bad things happen in threes. We argue that we are influenced by the stars, by the phases of the moon, by the motion of the planets in the heavens. When we get cancer, we automatically assume that something—or someone—is to blame.

But many events are not fully predictable or explicable. Disasters happen randomly, to good people as well as to bad ones, to star-crossed individuals as well as those who have a favorable planetary alignment. Sometimes you can make a good guess about the future, but randomness can confound even the most solid predictions. Don't be surprised when you're outlived by the overweight, cigar-smoking, speed-fiend motorcyclist down the block.

What's more, random events can mimic nonrandom ones. Even the most sophisticated scientists can have difficulty telling the difference between a real effect and a random fluke. Randomness can make placebos seem like miracle cures, or harmless compounds appear to be deadly poisons, and can even create subatomic particles out of nothing.

The Second Law of Randomness: Some Events Are Impossible to Predict

If you walk into a Las Vegas casino and observe the crowd gathered around the craps table, you'll probably see someone who thinks he's on a lucky streak. Because he's won several rolls in a row, his brain tells him he's going to keep winning, so he keeps gambling. You'll probably also see someone who's been losing. The loser's brain, like the winner's, tells him to keep gambling. Since he's been losing for so long, he thinks he's due for a stroke of luck; he won't walk away from the table, for fear of missing out.

Contrary to what our brains are telling us, there's no mystical force that imbues a winner with a streak of luck, nor is there a cosmic sense of justice that ensures that a loser's luck will turn around. The universe doesn't care one whit whether you've been winning or losing; each roll of the dice is just like every other.

No matter how much effort you put into observing how the dice have been behaving or how meticulously you have been watching for people who seem to have luck on their side, you get absolutely no information about what the next roll of a fair die will be. The outcome of a die roll is entirely independent of its history. And as a result, any scheme to gain some sort of advantage by observing the table is doomed to fail. Events like these—independent, purely random events—defy any attempts to find a pattern because there is none to be found.

Randomness provides an absolute block against human ingenuity; it means that our logic, our science, our capacity for reason can penetrate only so far in predicting the behavior of the cosmos. Whatever methods you try, whatever theory you create, whatever logic you use to predict the next roll of a fair die, there's always a 5/6 chance you are wrong. Always.

The Third Law of Randomness: Random Events Behave Predictably in Aggregate Even If They're Not Predictable Individually

Randomness is daunting; it sets limits where even the most sophisticated theories cannot go, shielding elements of nature from even our most determined inquiries. Nevertheless, to say that something is random is not equivalent to saying that we can't understand it. Far from it.

Randomness follows its own set of rules—rules that make the behavior of a random process understandable and predictable.

These rules state that even though a single random event might be completely unpredictable, a collection of independent random events is extremely predictable—and the larger the number of events, the more predictable they become. The law of large numbers is a mathematical theorem that dictates that repeated, independent random events converge with pinpoint accuracy upon a predictable average behavior. Another powerful mathematical tool, the central limit theorem, tells you exactly how far off that average a given collection of events is likely to be. With these tools, no matter how chaotic, how strange, a random behavior might be in the short run, we can turn that behavior into stable, accurate predictions in the long run.

The rules of randomness are so powerful that they have given physics some of its most sacrosanct and immutable laws. Though the atoms in a box full of gas are moving at random, their collective behavior is described by a simple set of deterministic equations. Even the laws of thermodynamics derive their power from the predictability of large numbers of random events; they are indisputable only because the rules of randomness are so absolute.

Paradoxically, the unpredictable behavior of random events has given us the predictions in which we are most confident.

Randomness in Music[*]

Donald E. Knuth

Patterns that are perfectly pure and mathematically exact have a strong aesthetic appeal, as advocated by Pythagoras and Plato and their innumerable intellectual descendants. Yet a bit of irregularity and unpredictability can make a pattern even more beautiful. I was reminded of this fact as I passed by two decorative walls while walking yesterday from my home to my office: One wall, newly built, tries to emulate the regular rectangular pattern of a grid, but it looks sterile and unattractive to my eyes; the other wall consists of natural stones that fit together only approximately yet form a harmonious unity that I find quite pleasing.

I noticed similar effects when I was experimenting years ago with the design of computer-generated typefaces for the printing of beautiful books. A design somehow "came to life" when it was not constrained to be rigidly consistent.[1]

Similar examples abound in the musical realm, as well as in the world of visual images. For example, I'm told that people who synthesize music electronically discovered long ago that rhythms are more exciting if they don't go exactly "1, 2, 3, 4" but rather miss the beat very slightly and become what a mathematician might call "$1 + \delta_1$, $2 + \delta_2$, $3 + \delta_3$, $4 + \delta_4$." Although the discrepancies δ mount to only a few milliseconds, positive or negative, they enliven the music significantly by comparison with the deadly and monotonous pulsing that you hear when the δs are entirely zero.

Singers and saxophone players know better than to hit the notes of a melody with perfect pitch.

[*] This article is based on an informal talk given to the Stanford Music Affiliates on May 9, 1990.

Furthermore, we can take liberties with the "ideal" notes themselves. In an essay called "Chance in artistic creation," published in 1894,[2] August Strindberg recounted the following experience:

> A musician whom I knew amused himself by tuning his piano haphazardly, without any rhyme or reason. Afterwards he played Beethoven's *Sonate Pathétique* by heart. It was an unbelievable delight to hear an old piece come back to life. How often had I previously heard this sonata, always the same way, never dreaming that it was capable of being developed further!

And the notion of planned imperfection is not limited to variations in the performance of a given composition; it extends also to the choices of notes that a composer writes down. The main purpose of my talk today is to describe a way by which you could build a simple machine that will produce *random harmonizations of any given melody.*

More precisely, I'll show you how to produce $2^n + 2^{n-1}$ different harmonizations of any n-note melody, all of which are pretty good. A machine can easily generate any one of them, chosen at random, when its user plays the melody on a keyboard with one hand.

The method I shall describe was taught to me in 1969 by David Kraehenbuehl (1923–1997), when I audited his class on keyboard harmony at Westminster Choir College. It is extremely simple, although you do need to understand the most elementary aspects of music theory. I shall assume that you are familiar with ordinary music notation.

Kraehenbuehl's algorithm produces four-part harmony from a given melody, where the top three parts form "triadic" chords and the bottom part supplies the corresponding bass notes.

A *triad* is a chord that consists of three notes separated by one-note gaps in the scale. Thus the triads are

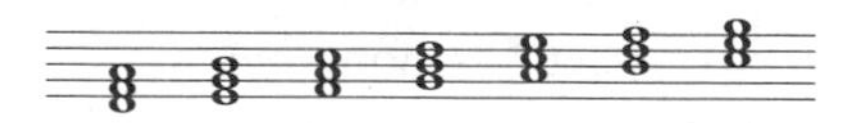

and others that differ only by being one or more octaves higher or lower. The bottom note of a triad is called its "root," and the other two notes are called its "third" and its "fifth."

These notions apply to any clef and to any key signature. For example, with the treble clef and in the key of G major, the seven triads

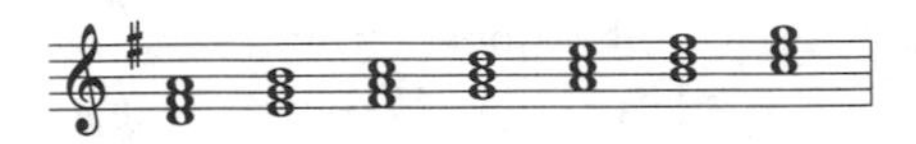

are known by more precise names such as a "D major triad," etc.; but we don't need to concern ourselves with such technical details.

The important thing for our purposes is to consider what happens when individual notes of a triad move up or down by an octave. If we view these chords modulo octave jumps, we see that they make a different shape on the staff when the root tone is moved up an octave so that the third tone becomes lowest; this change gives us the *first inversion* of the triad. And if the root and third are both moved up an octave, leaving the fifth tone lowest, we obtain the *second inversion:*

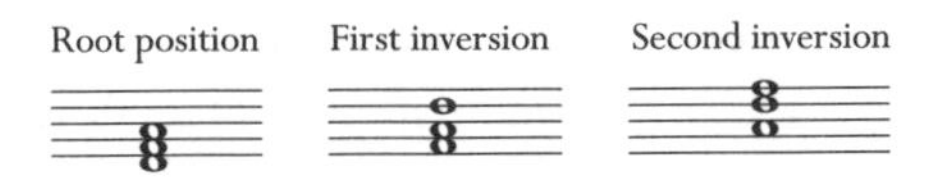

Even though two-note gaps appear between adjacent notes of the first and second inversions, these chords are still regarded as triads, because octave jumps don't change the name of a note on its scale: An A is still an A, etc., and no two notes of an inverted triad have adjacent names.

Music theorists have traditionally studied three-note chords by focusing their attention first on the root of each triad, and next on the bottom note, which identifies the inversion. Kraehenbuehl's innovation was to concentrate rather on the *top* note because it's the melody note.

He observed that each melody note in the scale comes at the top of three triadic chords—one in root position (0), one in first inversion (1), and one in second inversion (2):

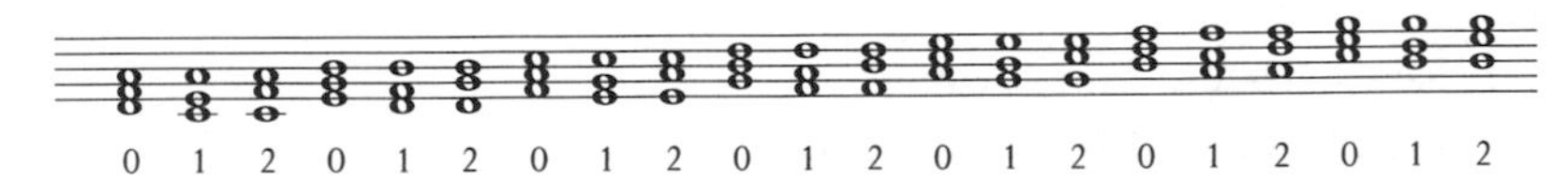

Furthermore, said Kraehenbuehl, there's a natural way to add a fourth part to this three-part harmony by simply repeating the root note an octave or two lower. For example, in the key of C, we get

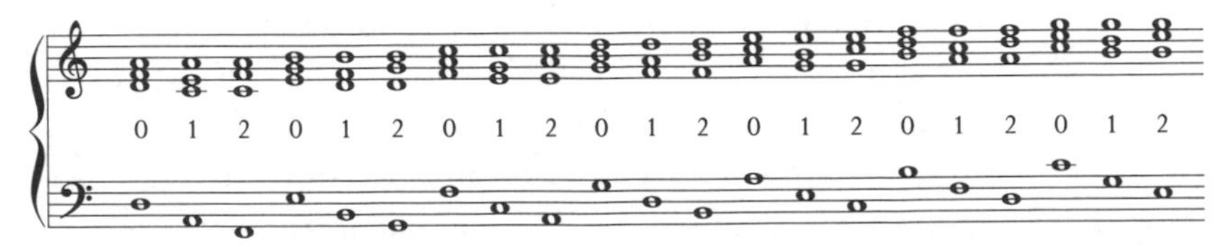

as the four-part harmonizations of melody notes A, A, A, B, B, B, . . . , G, G, G. This rule almost always works well; but like all good rules it has an exception: When the bass note turns out to be the *leading tone* of the scale (which is one below the tonic), we should change it to the so-called *dominant tone* (which is two notes lower). Thus Kraehenbuehl's correction to the natural rule yields the 21 four-part chords

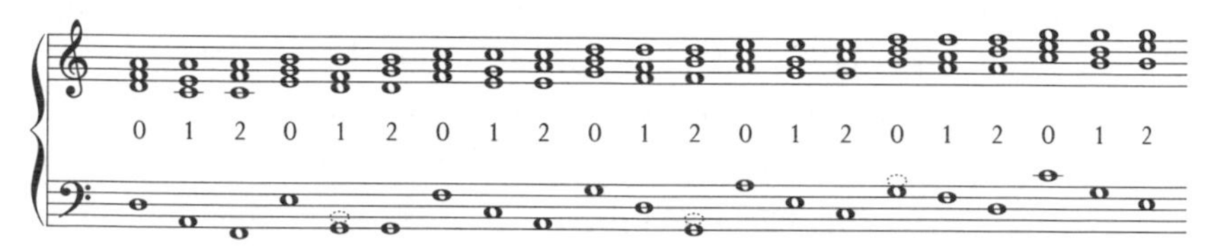

in the key of C, when B is the leading tone; and it yields

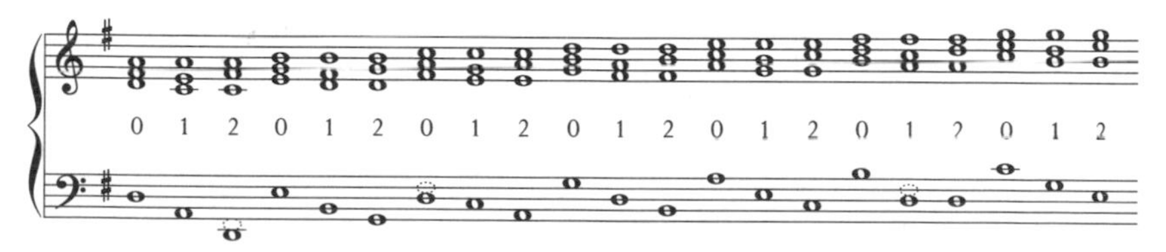

in the key of G, because F$^{\#}$ is the leading tone in that case. Notice that when the bass note is corrected by shifting it down from a leading tone in root position, it produces a chord with four separate pitches (the so-called "dominant seventh chord" of its key), so it's no longer a triad.

Okay, now we know three good ways to harmonize any given melody note in any given key. Kraehenbuehl completed his method by pointing out that the same principles apply to melodies with any number of notes, *provided only that we never use the same inversion twice in a row.* If one note has been harmonized with, say, the first inversion, the next note should be harmonized with either the root position or the second inversion; and so on. With this procedure there are three choices for the first chord, and two choices for every chord thereafter.

Let's test his algorithm by trying it out on a familiar melody:

"London Bridge is falling down, my fair lady" has eleven notes, so Kraehenbuehl has given us $3 \cdot 2 \cdot 2 \cdot 2 \cdot 2 \cdot 2 \cdot 2 \cdot 2 \cdot 2 \cdot 2 \cdot 2 = 3{,}072$ ways to harmonize it. The binary representations of three fundamental constants,

$$\pi = 3 + (0.0010010000111111011010101010001\ldots)_2,$$
$$e = 2 + (0.1011011111100001010100010110 0\ldots)_2,$$
$$\phi = 1 + (0.1001111000110111011110011 0111\ldots)_2,$$

serve to define three more-or-less random sequences of suitable inversions, if we prefer mathematical guidance to coin-flipping. Namely, we can use the integer part of the constant to specify the first chord, then we can change the number of inversions by $+1$ or -1 (modulo 3) for each successive binary digit that is 0 or 1, respectively. This procedure gives us three new harmonizations of that classic British theme:

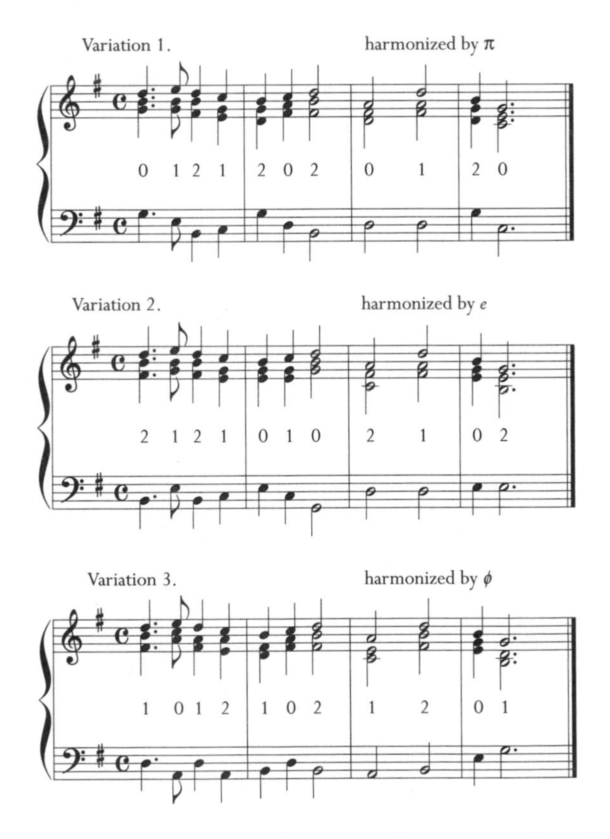

Amazing. Kraehenbuehl's algorithm seems far too simple to be correct, yet it really works!

Of course, there's a little glitch at the end, because we have only one chance in three of ending on a chord that's stable and "resolved." No problem: In such a case we can just repeat the last melody note. With this extension, variations 1 and 2 will end nicely, with

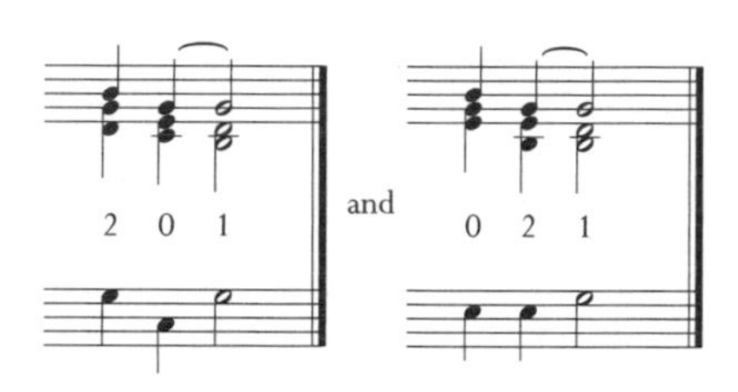

I can hardly wait for somebody to build me a keyboard that will perform such harmonizations automatically. After all, it's basically just a small matter of programming.

Notes

1. See, for example, my book *Digital Typography*, pages 57–59, 286–287, 324–325, 386, 391–396; also *The METAFONTbook*, pages 183–185.

2. August Strindberg, "Du Hasard dans la production artistique," *La Revue des revues* **11** (Nov. 15, 1894), 265–270.

Playing the Odds

Soren Johnson

One of the most powerful tools a designer can use when developing games is probability, using random chance to determine the outcome of player actions or to build the environment in which play occurs. The use of luck, however, is not without its pitfalls, and designers should be aware of the tradeoffs involved—what chance can add to the experience and when it can be counterproductive.

Failing at Probability

One challenge with using randomness is that humans are notoriously poor at evaluating probability accurately. A common example is the *gambler's fallacy,* which is the belief that odds even out over time. If the roulette wheel comes up black five times in a row, players often believe that the odds of it coming up black again are quite small, even though clearly the streak makes no difference whatsoever. Conversely, people also see streaks where none actually exist—the shooter with a "hot hand" in basketball, for example, is a myth. Studies show that, if anything, a successful shot actually predicts a subsequent miss.

Also, as designers of slot machines and massively multiplayer online games are quite aware, setting odds unevenly between each progressive reward level makes players think that the game is more generous than it really is. One commercial slot machine had its payout odds published by wizardofodds.com in 2008:

1:1 per 8 plays
2:1 per 600 plays
5:1 per 33 plays
20:1 per 2,320 plays

80:1 per 219 plays
150:1 per 6,241 plays

The 80:1 payoff is common enough to give players the thrill of beating the odds for a "big win" but still rare enough that the casino is at no risk of losing money. Furthermore, humans have a hard time estimating extreme odds—a 1 percent chance is anticipated too often, and 99 percent odds are considered to be as safe as 100 percent.

Leveling the Field

These difficulties in estimating odds accurately actually work in the favor of the game designer. Simple game-design systems, such as the dice-based resource-generation system in *Settlers of Catan,* can be tantalizingly difficult to master with a dash of probability.

In fact, luck makes a game more accessible because it shrinks the gap—whether in perception or in reality—between experts and novices. In a game with a strong luck element, beginners believe that no matter what, they have a chance to win. Few people would be willing to play a chess grand master, but playing a backgammon expert is much more appealing—a few lucky throws can give anyone a chance.

In the words of designer Dani Bunten, "Although most players hate the idea of random events that will destroy their nice safe predictable strategies, nothing keeps a game alive like a wrench in the works. Do not allow players to decide this issue. They don't know it but we're offering them an excuse for when they lose ('It was that damn random event that did me in!') and an opportunity to 'beat the odds' when they win."

Thus luck serves as a social lubricant—the alcohol of gaming, so to speak—that increases the appeal of multiplayer gaming to audiences that would not normally be suited for cutthroat head-to-head competition.

Where Luck Fails

Nonetheless, randomness is not appropriate for all situations or even all games. The "nasty surprise" mechanic is never a good idea. If a crate provides ammo and other bonuses when opened but explodes 1 percent of the time, the player has no chance to learn the probabilities in a safe manner. If the explosion occurs early enough, the player may stop opening crates immediately. If it happens much later, the player may feel unprepared and cheated.

Also, when randomness becomes just noise, the luck simply detracts from the player's understanding of the game. If a die roll is made every time a *StarCraft* Marine shoots at a target, the rate of fire simply appears uneven. Over time, the effect of luck on the game's outcome is negligible, but the player has a harder time grasping how strong a Marine's attack actually is with all the extra random noise.

Furthermore, luck can slow down a game unnecessarily. The board games *History of the World* and *Small World* have a similar conquest mechanic, except that the former uses dice and the latter does not (until the final attack). Making a die roll with each attack causes a *History of the World* turn to last at least three or four times as long as a turn in *Small World.* The reason is not just the logistical issues of rolling so many dice—knowing that the results of one's decisions are completely predictable allows one to plan out all the steps at once without worrying about contingencies. Often, handling contingencies is a core part of the game design, but game speed is an important factor too, so designers should be sure that the tradeoff is worthwhile.

Finally, luck is inappropriate for calculations to determine victory. Unlucky rolls feel the fairest the longer players are given to react to them before the game's end. Thus the earlier luck plays a role, the better for the perception of game balance. Many classic card games—pinochle, bridge, and hearts—follow a standard model of an initial random distribution of cards that establishes the game's "terrain," followed by a luck-free series of tricks that determines the winners and losers.

Probability Is Content

Indeed, the idea that randomness can provide an initial challenge to be overcome plays an important role in many classic games, from simple games such as *Minesweeper* to deeper ones such as *NetHack* and *Age of Empires.* At their core, solitaire and *Diablo* are not so different—both present a randomly generated environment that the player needs to navigate intelligently for success.

An interesting recent use of randomness is *Spelunky,* which is indie developer Derek Yu's combination of the random level generation of

NetHack with the game mechanics of 2-D platformers such as *Lode Runner.* The addictiveness of the game comes from the unlimited number of new caverns to explore, but frustration can emerge from the wild difficulty of certain unplanned combinations of monsters and tunnels.

In fact, pure randomness can be an untamed beast, creating game dynamics that throw an otherwise solid design out of balance. For example, *Civilization III* introduced the concept of strategic resources that were required to construct certain units—chariots need horses, tanks need oil, and so on. These resources were sprinkled randomly across the world, which inevitably led to large continents with only one cluster of iron controlled by a single artificial intelligence (AI) opponent. Complaints of being unable to field armies for lack of resources were common among the community.

For *Civilization IV,* the problem was solved by adding a minimum amount of space between certain important resources so that two sources of iron never could be within seven tiles of each other. The result was a still unpredictable arrangement of resources around the globe but without the clustering that could doom an unfortunate player. On the other hand, the game actively encouraged clustering for less important luxury resources—incense, gems, and spices—to promote interesting trade dynamics.

Showing the Odds

Ultimately, when considering the role of probability, designers need to ask themselves, "How is luck helping or hurting the game?" Is randomness keeping the players pleasantly off balance so that they can't solve the game trivially? Or is it making the experience frustratingly unpredictable so that players are not invested in their decisions?

One factor that helps to ensure the former is making the probability as explicit as possible. The strategy game *Armageddon Empires* based combat on a few simple die rolls and then *showed* the dice directly onscreen. Allowing the players to peer into the game's calculations increases their comfort level with the mechanics, which makes chance a tool for the player instead of a mystery.

Similarly, with *Civilization IV,* we introduced a help mode that showed the exact probability of success in combat, which drastically increased player satisfaction with the underlying mechanics. Because humans have such a hard time estimating probability accurately, helping them make a smart decision can improve the experience immensely.

Some deck-building card games, such as *Magic: The Gathering* or *Dominion,* put probability in the foreground by centering the game experience on the likelihood of drawing cards in the player's carefully constructed deck. These games are won by players who understand the proper ratio of rares to commons, knowing that each card is drawn exactly once each time through the deck. This concept can be extended to other games of chance by providing, for example, a virtual "deck of dice" that ensures that the distribution of die rolls is exactly even.

Another interesting—and perhaps underused—idea from the distant past of gaming history is the *element-of-chance* game option from the turn-based strategy game *Lords of Conquest.* The three options available—low, medium, and high—determined whether luck was used only to break ties or to play a larger role in resolving combat. The appropriate role of chance in a game ultimately is a subjective question, and giving players the ability to adjust the knobs themselves can open up the game to a larger audience with a greater variety of tastes.

Machines of the Infinite

John Pavlus

On a snowy day in Princeton, New Jersey, in March 1956, a short, owlish-looking man named Kurt Gödel wrote his last letter to a dying friend. Gödel addressed John von Neumann formally even though the two had known each other for decades as colleagues at the Institute for Advanced Study in Princeton. Both men were mathematical geniuses, instrumental in establishing the U.S. scientific and military supremacy in the years after World War II. Now, however, von Neumann had cancer, and there was little that even a genius like Gödel could do except express a few overoptimistic pleasantries and then change the subject:

> Dear Mr. von Neumann:
>
> With the greatest sorrow I have learned of your illness. . . . As I hear, in the last months you have undergone a radical treatment and I am happy that this treatment was successful as desired, and that you are now doing better. . . .
>
> Since you now, as I hear, are feeling stronger, I would like to allow myself to write you about a mathematical problem, of which your opinion would very much interest me. . . .

Gödel's description of this problem is utterly unintelligible to nonmathematicians. (Indeed, he may simply have been trying to take von Neumann's mind off of his illness by engaging in an acutely specialized version of small talk.) He wondered how long it would take for a hypothetical machine to spit out answers to a problem. What he concluded sounds like something out of science fiction:

> If there really were [such] a machine . . . this would have consequences of the greatest importance. Namely, it would obviously mean that . . . the mental work of a mathematician concerning Yes-or-No questions could be completely replaced by a machine.

By "mental work," Gödel didn't mean trivial calculations like adding 2 and 2. He was talking about the intuitive leaps that mathematicians take to illuminate entirely new areas of knowledge. Twenty-five years earlier, Gödel's now famous incompleteness theorems had forever transformed mathematics. Could a machine be made to churn out similar world-changing insights on demand?

A few weeks after Gödel sent his letter, von Neumann checked into Walter Reed Army Medical Center in Washington, D.C., where he died less than a year later, never having answered his friend. But the problem would outlive both of them. Now known as P versus NP, Gödel's question went on to become an organizing principle of modern computer science. It has spawned an entirely new area of research called computational complexity theory—a fusion of mathematics, science, and engineering that seeks to prove, with total certainty, what computers can and cannot do under realistic conditions.

But P versus NP is about much more than just the plastic-and-silicon contraptions we call computers. The problem has practical implications for physics and molecular biology, cryptography, national security, evolution, the limits of mathematics, and perhaps even the nature of reality. This one question sets the boundaries for what, in theory, we will ever be able to compute. And in the 21st century, the limits of computation look more and more like the limits of human knowledge itself.

The Bet

Michael Sipser was only a graduate student, but he knew someone would solve the P versus NP problem soon. He even thought he might be the one to do it. It was the fall of 1975, and he was discussing the problem with Leonard Adleman, a fellow graduate student in the computer science department at the University of California, Berkeley. "I had a fascination with P versus NP, had this feeling that I was somehow able to understand it in a way that went beyond the way everyone else seemed to be approaching it," says Sipser, who is now head of the mathematics department at the Massachusetts Institute of Technology. He was so sure of himself that he made a wager that day with Adleman: P versus NP would be solved by the end of the 20th century, if not sooner. The terms: one ounce of pure gold.

Sipser's bet made a kind of poetic sense because P versus NP is itself a problem about how quickly other problems can be solved. Sometimes simply following a checklist of steps gets you to the end result in relatively short order. Think of grocery shopping: You tick off the items one by one until you reach the end of the list. Complexity theorists label these problems P, for "polynomial time," which is a mathematically precise way of saying that no matter how long the grocery list becomes, the amount of time that it will take to tick off all the items will never grow at an unmanageable rate.

In contrast, many more problems may or may not be practical to solve by simply ticking off items on a list, but checking the solution is easy. A jigsaw puzzle is a good example: Even though it may take effort to put together, you can recognize the right solution just by looking at it. Complexity theorists call these quickly checkable, "jigsaw puzzle-like" problems NP.

Four years before Sipser made his bet, a mathematician named Stephen Cook had proved that these two kinds of problems are related: Every quickly solvable P problem is also a quickly checkable NP problem. The P versus NP question that emerged from Cook's insight—and that has hung over the field ever since—asks if the reverse is also true: Are all quickly checkable problems quickly solvable as well? Intuitively speaking, the answer seems to be no. Recognizing a solved jigsaw puzzle ("Hey, you got it!") is hardly the same thing as doing all the work to find the solution. In other words, P does not seem to equal NP.

What fascinated Sipser was that nobody had been able to mathematically *prove* this seemingly obvious observation. And without a proof, a chance remained, however unlikely or strange, that all NP problems might actually be P problems in disguise. P and NP might be equal—and because computers can make short work of any problem in P, P equals NP would imply that computers' problem-solving powers are vastly greater than we ever imagined. They would be exactly what Gödel described in his letter to von Neumann: mechanical oracles that could efficiently answer just about any question put to them, so long as they could be programmed to verify the solution.

Sipser knew that this outcome was vanishingly improbable. Yet proving the opposite, much likelier, case—that P is not equal to NP—would be just as groundbreaking.

Like Gödel's incompleteness theorems, which revealed that mathematics must contain true but unprovable propositions, a proof showing that P does not equal NP would expose an objective truth concerning the limitations of knowledge. Solving a jigsaw puzzle and recognizing that one is solved are two fundamentally different things, and there are no shortcuts to knowledge, no matter how powerful our computers get.

Proving a negative is always difficult, but Gödel had done it. So to Sipser, making his bet with Adleman, 25 years seemed like more than enough time to get the job done. If he couldn't prove that P did not equal NP himself, someone else would. And he would still be one ounce of gold richer.

Complicated Fast

Adleman shared Sipser's fascination, if not his confidence, because of one cryptic mathematical clue. Cook's paper establishing that P problems are all NP had also proved the existence of a special kind of quickly checkable type of problem called NP-complete. These problems act like a set of magic keys: If you find a fast algorithm for solving one of them, that algorithm can also unlock the solution to every other NP problem and prove that P equals NP.

There was just one catch: NP-complete problems are among the hardest anyone in computer science had ever seen. And once discovered, they began turning up everywhere. Soon after Cook's paper appeared, one of Adleman's mentors at Berkeley, Richard M. Karp, published a landmark study showing that 21 classic computational problems were all NP-complete. Dozens, then hundreds, soon followed. "It was like pulling a finger out of a dike," Adleman says. Scheduling air travel, packing moving boxes into a truck, solving a Sudoku puzzle, designing a computer chip, seating guests at a wedding reception, playing Tetris, and thousands of other practical, real-world problems have been proved to be NP-complete.

How could this tantalizing key to solving P versus NP seem so commonplace and so uncrackable at the same time? "That's why I was interested in studying the P versus NP problem," says Adleman, who is now a professor at the University of Southern California. "The power and breadth of these computational questions just seemed deeply awesome.

The Basics of Complexity

How long will it take to solve that problem? That's the question that researchers ask as they classify problems into computational classes. As an example, consider a simple sorting task: Put a list of random numbers in order from smallest to largest. As the list gets bigger, the time it takes to sort the list increases at a manageable rate—as the square of the size of the list, perhaps. This puts it in class "P" because it can be solved in polynomial time. Harder questions, such as the "traveling salesman" problem [*see next box in this article*], require exponentially more time to solve as they grow more complex. These "NP-complete" problems quickly get so unwieldy that not even billions of processors working for billions of years can crack them.

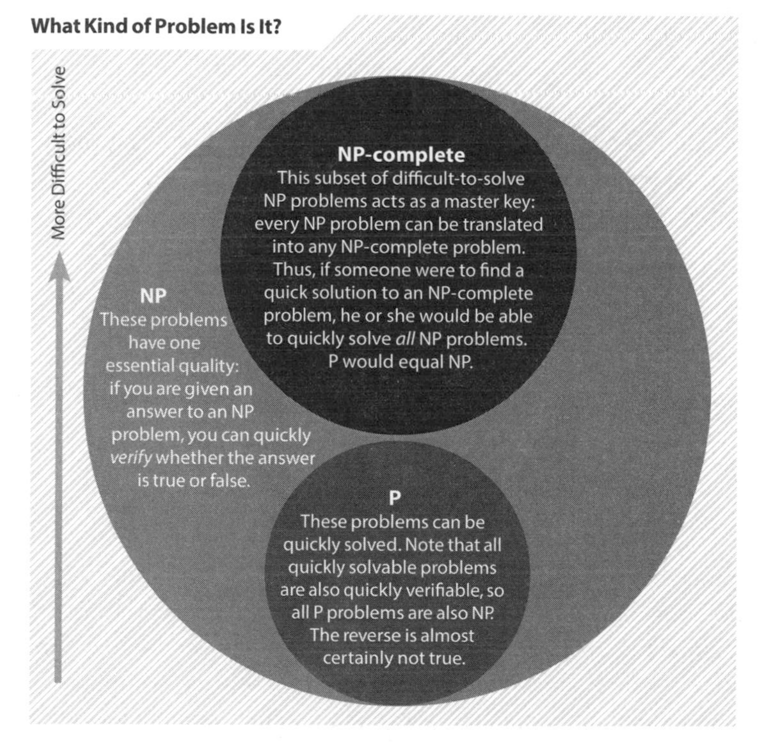

Venn diagram. Reproduced with permission. Copyright © 2012 Scientific American, a division of Nature America, Inc. All Rights Reserved.

But we certainly didn't understand them. And it didn't seem like we would be understanding them anytime soon." (Adleman's pessimism about P versus NP led to a world-changing invention: A few years after making his bet, Adleman and his colleagues Ronald Rivest and Adi Shamir exploited the seeming incommensurability of P and NP to create their eponymous RSA encryption algorithm, which remains in wide use for online banking, communications, and national security applications.)

NP-complete problems are hard because they get complicated fast. Imagine that you are a backpacker planning a trip through a number of cities in Europe, and you want a route that takes you through each city while minimizing the total distance you will need to travel. How do you find the best route? The simplest method is just to try out each possibility. With five cities to visit, you need to check only 12 possible routes. With 10 cities, the number of possible routes mushrooms to more than 180,000. At 60 cities, the number of paths exceeds the number of atoms in the known universe. This computational nightmare is known as the traveling salesman problem, and in more than 80 years of intense study, no one has ever found a general way to solve it that works better than trying every possibility one at a time.

That is the perverse essence of NP-completeness—and of P versus NP: Not only are all NP-complete problems equally impossible to solve except in the simplest cases—even if your computer has more memory than God and the entire lifetime of the universe to work with—they seem to pop up everywhere. In fact, these NP-complete problems don't just frustrate computer scientists. They seem to put limits on the capabilities of nature itself.

Nature's Code

The pioneering Dutch programmer Edsger Dijkstra understood that computational questions have implications beyond mathematics. He once remarked that "computer science is no more about computers than astronomy is about telescopes." In other words, computation is a behavior exhibited by many systems besides those made by Google and Intel. Indeed, any system that transforms inputs into outputs by a set of discrete rules—including those studied by biologists and physicists—can be said to be computing.

The Swedish Salesman

If you're feeling ambitious on your next trip to Sweden, consider seeing it all. Researchers have proved that the route pictured here is the shortest possible path that crosses through every one of the country's 24,978 cities, towns, and villages. Researchers don't expect anyone to make the actual trip, but the search techniques they developed to solve it will help in other situations where investigators need to find the optimal path through a complicated landscape—in microchip design or genome sequencing, for instance. Image courtesy of William J. Cook.

In 1994 mathematician Peter Shor proved that cleverly arranged subatomic particles could break modern encryption schemes. In 2002 Adleman used strands of DNA to find an optimal solution to an instance of the traveling salesman problem. And in 2005 Scott Aaronson, an expert in quantum computing who is now at MIT's Computer Science and Artificial Intelligence Laboratory, used soap bubbles, of all things, to efficiently compute optimal solutions to a problem known as the Steiner tree. These are all exactly the kinds of NP problems on which computers should choke their circuit boards. Do these natural systems know something about P versus NP that computers don't?

"Of course not," Aaronson says. His soap bubble experiment was actually a reductio ad absurdum of the claim that simple physical systems can somehow transcend the differences between P and NP problems. Although the soap bubbles did "compute" perfect solutions to the minimum Steiner tree in a few instances, they quickly failed as the size of the problem increased, just as a computer would. Adleman's DNA-strand experiment hit the same wall. Shor's quantum algorithm does work in all instances, but the factoring problem that it cracks is almost certainly not NP-complete. Therefore, the algorithm doesn't provide the key that would unlock every other NP problem. Biology, classical physics, and quantum systems all seem to support the idea that NP-complete problems have no shortcuts. And that would only be true if P did not equal NP.

"Of course, we still can't prove it with airtight certainty," Aaronson says. "But if we were physicists instead of complexity theorists, 'P does not equal NP' would have been declared a law of nature long ago—just like the fact that nothing can go faster than the speed of light." Indeed, some physical theories about the fundamental nature of the universe—such as the holographic principle, suggested by Stephen Hawking's work on black holes—imply that the fabric of reality itself is not continuous but made of discrete bits, just like a computer (see "Is Space Digital?" by Michael Moyer, *Scientific American*, February 2012). Therefore, the apparent intractability of NP problems—and the limitations on knowledge that this implies—may be baked into the universe at the most fundamental level.

Brain Machine

So if the very universe itself is beholden to the computational limits imposed by P versus NP, how can it be that NP-complete problems seem to get solved all the time—even in instances where finding these solutions should take trillions of years or more?

For example, as a human fetus gestates in the womb, its brain wires itself up out of billions of individual neurons. Finding the best arrangement of these cells is an NP-complete problem—one that evolution appears to have solved. "When a neuron reaches out from one point to get to a whole bunch of other synapse points, it's basically a graph-optimization problem, which is NP-hard," says evolutionary neurobiologist Mark Changizi. Yet the brain doesn't actually solve the problem—it makes a close approximation. (In practice, the neurons consistently get within 3 percent of the optimal arrangement.) The *Caenorhabditis elegans* worm, which has only 302 neurons, still doesn't have a perfectly optimal neural-wiring diagram, despite billions on billions of generations of natural selection acting on the problem. "Evolution is constrained by P versus NP," Changizi says, "but it works anyway because life doesn't always require perfection to function well."

And neither, it turns out, do computers. That modern computers can do anything useful at all—much less achieve the wondrous feats we all take for granted on our video-game consoles and smartphones—is proof that the problems in P encompass a great many of our computing needs. For the rest, often an imperfect approximating algorithm is good enough. In fact, these "good enough" algorithms can solve immensely complex search and pattern-matching problems, many of which are technically NP-complete. These solutions are not always mathematically optimal in every case, but that doesn't mean they aren't useful.

Take Google, for instance. Many complexity researchers consider NP problems to be, in essence, search problems. But according to Google's director of research Peter Norvig, the company takes pains to avoid dealing with NP problems altogether. "Our users care about speed more than perfection," he says. Instead Google researchers optimize their algorithms for an even faster computational complexity category than P (referred to as linear time) so that search results appear nearly instantaneously. And if a problem comes up that cannot be solved in this way? "We either reframe it to be easier, or we don't bother," Norvig says.

That is the legacy and the irony of P versus NP. Writing to von Neumann in 1956, Gödel thought the problem held the promise of a future filled with infallible reasoning machines capable of replacing "the mental work of a mathematician" and churning out bold new truths at the push of a button. Instead decades of studying P versus NP have helped build a world in which we extend our machines' problem-solving powers by embracing their limitations. Lifelike approximation, not mechanical perfection, is how Google's autonomous cars can drive themselves on crowded Las Vegas freeways and IBM's Watson can guess its way to victory on *Jeopardy*.

Gold Rush

The year 2000 came and went, and Sipser mailed Adleman his ounce of gold. "I think he wanted it to be embedded in a cube of Lucite, so he could put it on his desk or something," Sipser says. "I didn't do that." That same year the Clay Mathematics Institute in Cambridge, Mass., offered a new bounty for solving P versus NP: $1 million. The prize helped to raise the problem's profile, but it also attracted the attention of amateurs and cranks; nowadays, like many prominent complexity theorists, Sipser says, he regularly receives unsolicited e-mails asking him to review some new attempt to prove that P does not equal NP—or worse, the opposite.

Although P versus NP remains unsolved, many complexity researchers still think it will yield someday. "I never really gave up on it," Sipser says. He claims to still pull out pencil and paper from time to time and work on it—almost for recreation, like a dog chewing on a favorite bone. P versus NP is, after all, an NP problem itself: The only way to find the answer is to keep searching. And while that answer may never come, if it does, we will know it when we see it.

More to Explore

The Efficiency of Algorithms. Harry R. Lewis and Christos H. Papadimitriou in *Scientific American,* Vol. 238, No. 1, pp 96–109, January 1978.

"The History and Status of the P versus NP Question." Michael Sipser in *Proceedings of the Twenty-Fourth Annual ACM Symposium on Theory of Computing,* pp 603–618, 1992.

"The Limits of Reason." Gregory Chaitin in *Scientific American,* Vol. 294, No. 3, pp 74–81, March 2006.

"The Limits of Quantum Computers." Scott Aaronson in *Scientific American,* Vol. 298, No. 3, pp 62–69, March 2008.

Bridges, String Art, and Bézier Curves

Renan Gross

The Jerusalem Chords Bridge

The Jerusalem Chords Bridge, in Israel, was built to make way for the city's light rail train system. However, its design took into consideration more than just utility—it is a work of art, designed as a monument. Its beauty rests not only in the visual appearance of its criss-cross cables, but also in the mathematics that lies behind it. Let us take a deeper look into these chords.

The Jerusalem Chords Bridge is a suspension bridge, which means that its entire weight is held from above. In this case, the deck is connected to a single tower by powerful steel cables. The cables are connected in the following way: The ones at the top of the tower support the center of the bridge, and the ones at the bottom support the further away sections, so that the cables cross each other.

Despite the fact that they draw out discrete, straight lines, we notice a remarkable feature: The outline of the cables' edges seems strikingly smooth. Does it obey any known mathematical formula?

To find out the shape that the edges make, we have to devise a mathematical model for the bridge. As the bridge itself is quite complex, featuring a curved deck and a two-part leaning tower, we have to simplify things. Although we lose a little accuracy and precision, we gain in mathematical simplicity, and we still capture the beautiful essence of the bridge's form. Afterward, we will be able to generalize our simple description and apply it to the real bridge structure.

This is the core of modeling—taking only the important features from the real world and translating them into mathematics.

Figure 1. The Jerusalem Chords Bridge at night. Image: Petdad [7].

Figure 2. The Jerusalem Chords Bridge as seen from below.

Chord Analysis

Let's look at a coordinate system, (x,y). The x axis corresponds to the base of the bridge, and the y axis to the tower from which it hangs.

Taking the tower to span from 0 to 1 on our y axis, and the deck from 0 to 1 on the x axis, we make n marks uniformly spaced on each axis. From each mark on the x axis, we draw a straight line to the y axis, so that the first mark on the x axis is connected to the nth on the y axis, the second on the x axis to the $(n - 1)$st on the y axis and so on. These lines represent our chords. Let's also assume that the x and y axes meet at a right angle. This is not a perfect picture of reality, for the cables are not evenly spaced and the tower and deck are not perpendicular, but it simplifies things.

The outline formed by the chords is essentially made out of the intersections of one cable and the one adjacent to it: You connect each intersection point with the one after it by a straight line. The more chords there are, the smoother the outline becomes. So the smooth curve that is hinted at by the chords, the *intersection envelope,* is the outline you would get from infinitely many chords.

Forgoing the detailed calculations (which you can find at *Plus* Internet magazine [8]), we find that all the points on this curve have coordinates of the type

$$(x,y) = (t^2, (1 - t)^2)$$

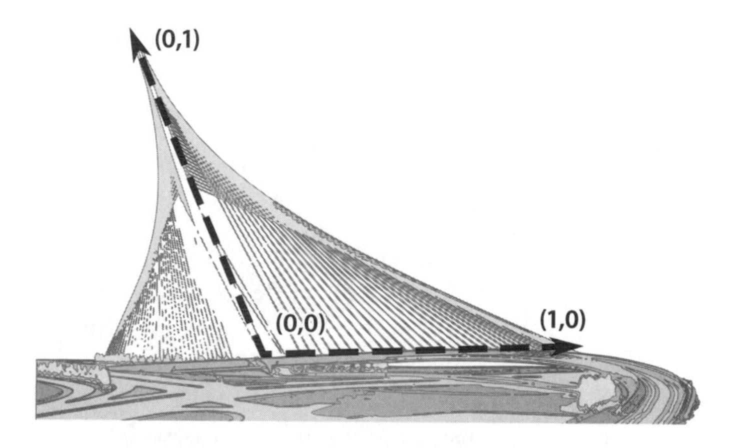

FIGURE 3. Coordinate axes superimposed on the bridge.

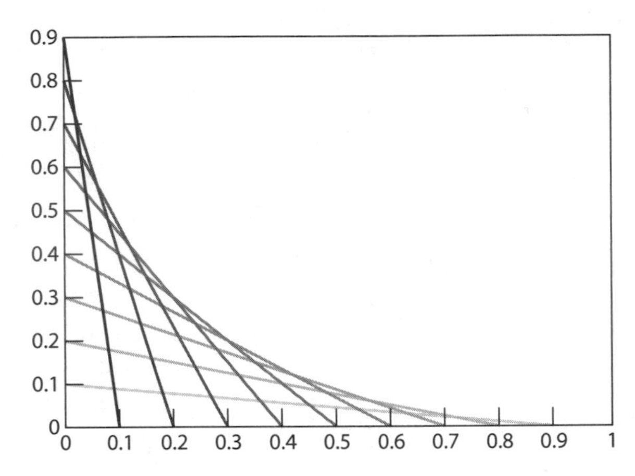

FIGURE 4. Our axes with evenly spaced chords.

for all t between 0 and 1. Hence,

$$y(x) = (1 - \sqrt{x})^2$$

"Well, is this it?" we ask. Is the shape going to remain an unnamed mathematical relation? In fact, no! Though it is not easy to see at first, this is actually the equation for a parabola! To this you might reply that the equation for a parabola is this:

$$y(x) = ax^2 + bx + c$$

which differs vastly from our result, and you would be correct. However, with a little work it can be shown that if we define $R = x + y$ and $S = x - y$, we can rewrite our unfamiliar equation as

$$R(S) = \frac{S^2}{2} + \frac{1}{2}$$

(See [9] for the details.)

And this indeed conforms to our well-known parabola equation. By replacing our variables x and y with S and R, we have actually rotated our coordinate system by 45 degrees. But this new coordinate system needn't frighten us. As we can see, the parabola equation is all the same.

The result we got—that the outline of the cables is essentially parabolic—is certainly satisfying, for the parabola is such a simple and elegant shape. But it also leaves us a bit puzzled. Is there a reason for

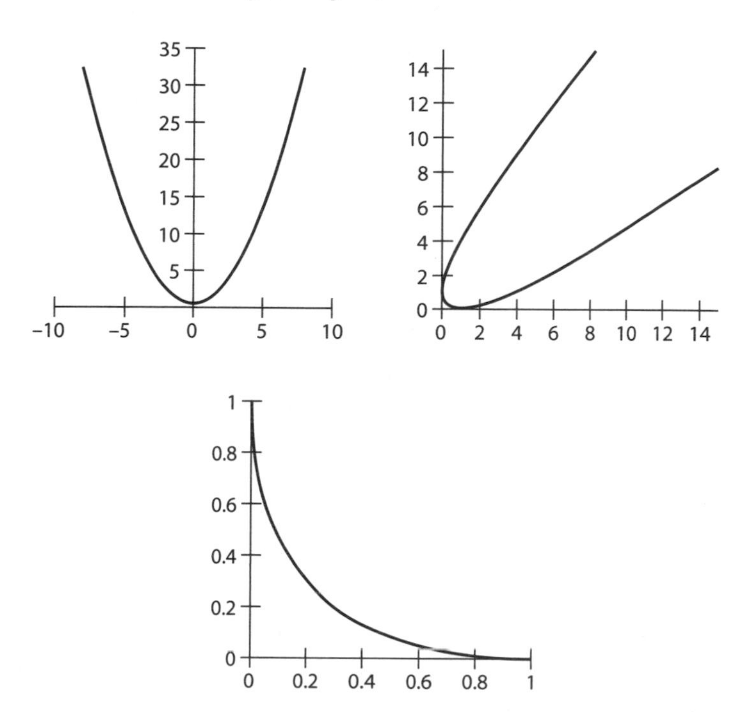

FIGURE 5. Top left: The parabola $R(S) = S^2/2 + 1/2$. Top right: The same parabola tilted by 45 degrees. Bottom: Zoom on the square defined by x and y ranging from 0 to 1 in the tilted parabola, which is the region that represents the bridge.

this simplicity? Why should it be parabolic, and not some other curve? If we changed the shape of the bridge—say, to make the tower more leaning—how would it affect our curve? Is there any way in which we can make amends for our simplifications and assumptions that we had to perform earlier?

An Unlikely Answer

The answers to our questions originate from a surprising field: that of automobile design. Back in the 1960s, engineer Pierre Bézier [10] used special curves to specify how he wanted car parts to look. These curves are called Bézier curves. We shall now take a look at what they have to offer us.

We all know that between any two points there can be only one straight line; hence, we can define a specific line using only two points. In a similar fashion, a Bézier curve is defined by any number of points, called *control points.* Unlike a straight line, it does not pass through all of the points. Rather, it starts at the first point and ends at the last, but it does not necessarily go through all the others. Instead, the points act as "weights," which direct the flow of the curve from the initial point to the last.

The number of points is used to define what is called the *degree* of the curve. A two-point *linear* Bézier curve has degree 1 and is just an ordinary straight line; a three-point *quadratic* Bézier curve has degree 2 and is a parabola; and in general, a curve of degree n has $n + 1$ control points.

A nice way of visualizing the construction of a Bézier curve is to imagine a pencil that starts drawing from the first control point to the last. On the way, it is attracted to the various control points, but the level of attraction changes as the pencil goes along. It is initially most attracted to the first control points, so as the pencil starts drawing, it heads off in their direction. As it progresses, it becomes more and more attracted to the later control points, until it finally reaches the last point. At any given time while we draw, we can ask, "What percentage of the curve has the pencil drawn already?" This percentage is called *the curve parameter* and is marked by t.

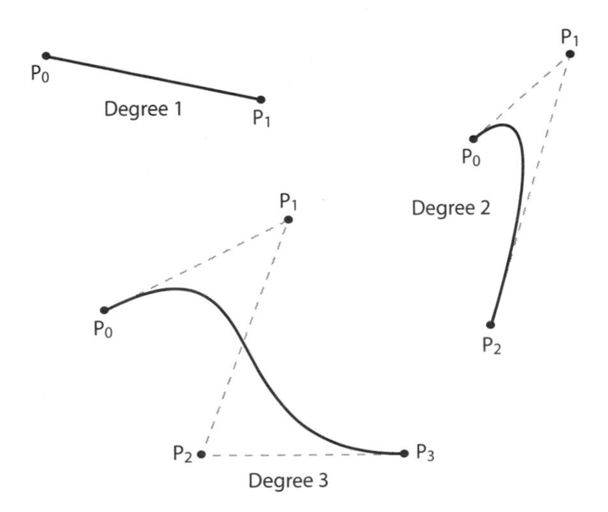

FIGURE 6. Bézier curves with degrees 1, 2, and 3.

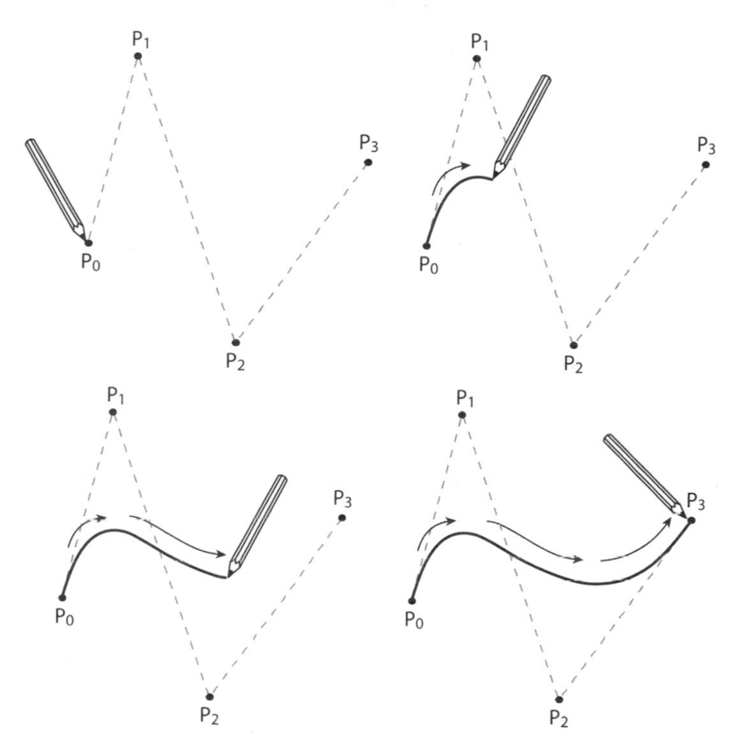

FIGURE 7. Drawing a Bézier curve.

How does all of this relate to our parabolic bridge? The connection is revealed when we take a look at how to actually draw a Bézier curve. One way of drawing it is to follow a mathematical formula that gives the coordinates of the curve. We will skip over that (you can have a look at the formula at [11]) and instead move on to the second method: constructing a Bézier curve recursively. In this method, to construct an nth degree curve, we use two $(n - 1)$st degree curves. It is best to illustrate this method with an example.

Suppose we have a 3rd degree curve. It is defined by 4 points: P_0, P_1, P_2, and P_3. From these we create two new groups: all points except the last, and all points except the first. We now have

Group 1: P_0, P_1, and P_2
Group 2: P_1, P_2, and P_3

Each of these groups defines a 2nd degree Bézier curve. Remember how we talked about using a pencil that moves from the first point to the last? Now, suppose that you have two pencils, and you draw both of these 2nd degree curves at the same time. The first curve starts from P_0 and finishes at P_2, and the second starts at P_1 and ends at P_3. At any given time during the pencils' journeys, you can connect their positions with a straight line.

So, while these two are drawing, think of a third pencil. This pencil is always somewhere on the line connecting the two current positions of the other pencils, and it moves along at the same rate as the other two. At the start, it is on the line connecting P_0 and P_1, and since both pencils have moved along 0% of their curves, the third pencil has moved along 0% of this line, which puts it at P_0. After the other two pencils have moved along, say, 36% of their curves and are at points Q_0 and Q_1, the third pencil is on the line from Q_0 to Q_1, at the point that marks 36% of that line. When the other two pencils have finished their journey, so they are at points P_2 and P_3 and have traveled 100% of the way, the third pencil is on the line from P_2 to P_3 at 100% along the way, which puts it at P_3.

There is a very good question to be asked here. We have just described how to create an nth degree curve, but doing so requires

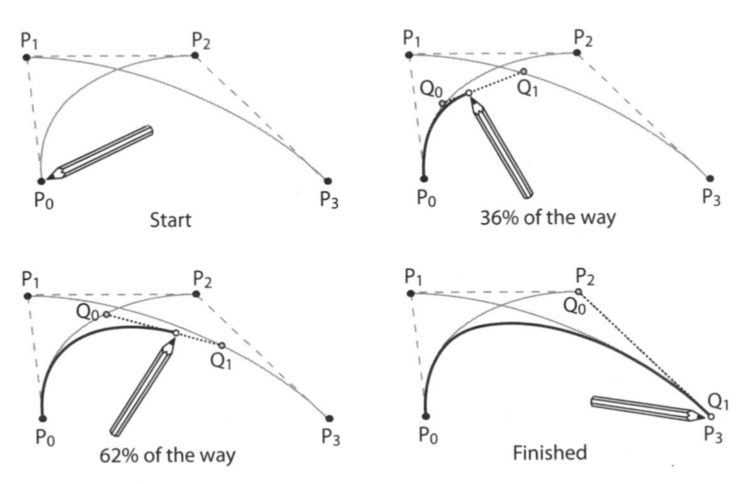

FIGURE 8. Building a cubic Bézier curve using quadratic curves. The P_0 to P_2 and P_1 to P_3 curves are the second-degree quadratic curves, whereas the P_0 to Q_1 curve is the third-degree cubic. This is the curve we want to construct. The points Q_0 and Q_1 go along the two second-degree curves. Our drawing pencil always goes along the line connecting Q_0 and Q_1.

drawing $(n-1)$st degree curves. How do we know how to do that? Luckily, we can apply exactly the same process to these assistant curves as well. We can build them out of two lesser degree curves. By repeating this process, we eventually reach a curve that we do know how to draw. This is the linear, 1st degree curve—it is just a straight line, which we have no problem drawing at all. Thus, all complex Bézier curves can be drawn using a composition of many straight lines.

Applying Bézier Curves

Now that we are a bit familiar with Bézier curves, we can return to our original questions: What made the bridge's shape come out parabolic? How can we extend our model to fix the assumptions we made?

It turns out that our beautiful Jerusalem Chords Bridge is nothing but a quadratic Bézier curve! To see this, let's go back to our coordinate system representing the bridge and draw a quadratic Bézier curve with the control points $P_0 = (0,1)$, $P_1 = (0,0)$, and $P_2 = (1,0)$. Using our recursive process, the quadratic curve is formed from two straight lines: the line from P_0 to P_1 (the y axis from 1 down to 0) and from P_1 to P_2 (the x axis from 0 up to 1).

Now suppose that the first pencil has traveled down the y axis by a distance t to the point $(0,1-t)$. In the same time, the second pencil has traveled along the x axis to the point $(0,t)$. The third pencil is therefore on the line L_t from $(0,1-t)$ to $(t,0)$, $100 \times t\%$ along the way. Thus, the Bézier curve meets all of the lines L_t for t between 0 and 1. These lines (or at least n of them) correspond to our bridge chords.

Now setting $t = 0.5$, we see that the halfway point $P = (0.25,0.25)$ of the line $L_{0.5}$, lies on our Bézier curve. Figure 10 shows that P also lies on the outline formed by the lines L_t. (See [12] if you're not convinced by the picture). This method is enough to show that the Bézier curve and the outline are one and the same curve. As Figure 10 shows, any other parabolic curve that meets all the L_t misses the point P and crosses the line $L_{0.5}$ twice.

Appreciating this fact allows us to deal with some of the previous model's inaccuracies. First, we assumed that the axes were perpendicular to each other, even though the tower and the deck of the bridge are actually at an angle. Now we see that this angle does not matter. The argument we just used continues to hold if we increase the angle

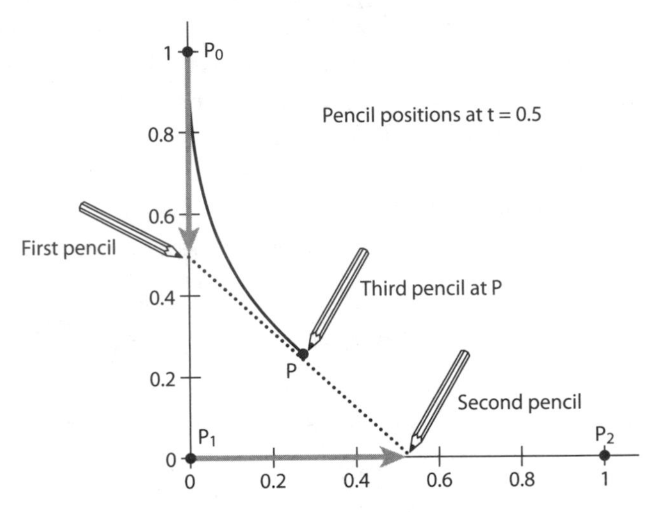

FIGURE 9. The arrows inside the x and y axes represent the distance t traveled along the axes for $t = 0.5$. The diagonal line is the line $L_{0.5}$.

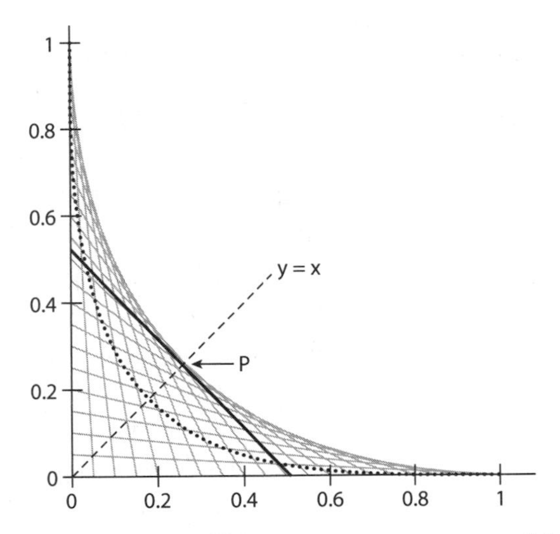

FIGURE 10. The lines that resemble netting represent some of the lines L_t. The diagonal line is $L_{0.5}$. The parabola inside the netting illustrates that any parabolic shape apart from the outline curve misses the point P.

between the axes by rotating the y axis (and any other line radiating out from the point (0,0) in between the x and y axes) counterclockwise by the necessary amount. We know that any quadratic Bézier curve is a parabola, so the bridge outline is still a parabola. Second, we see that it does not matter whether the chords in our model are evenly spaced

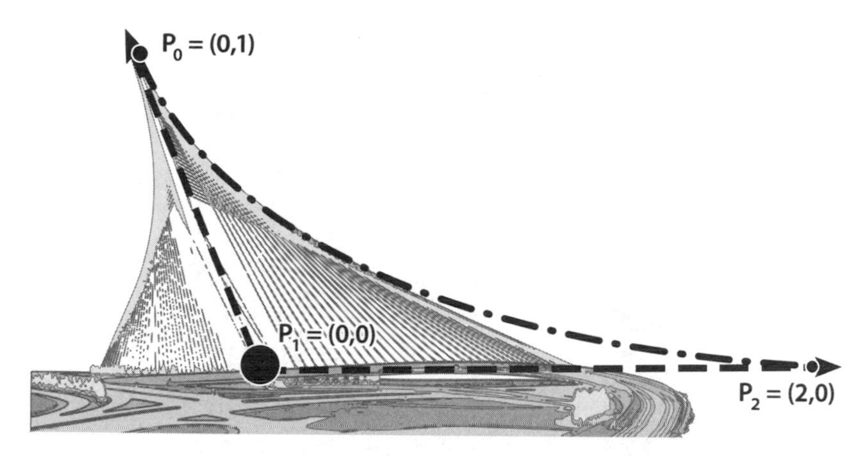

FIGURE 11. The correct Bézier curve applied to the Jerusalem Chords Bridge.

or not: They just represent some of the straight lines L_t whose outlines define our Bézier curve. Third, the chords coming out of the tower do not span its entire length, but stop about halfway. This just means that we have to look at a partial Bézier curve. If we extend the deck onward, we still have a parabola, with P_2 as (2,0). Then the outline of the chords is just the portion of the parabola that spans until (0,1).

We can now rest peacefully, knowing the underlying reason for the parabolic shape of the Jerusalem Chords Bridge outline. Somehow, a curve that was used in the 1960s for designing car parts managed to sneak its way into 21st century bridges!

Béziers Everywhere!

The abundance of uses for Bézier curves is far greater than just for cars and bridges. It finds its way into many more fields and applications. One such field is that of string art, in which strings are spread across a board filled with nails. Although the strings can only make straight lines, a great many of them at different angles can generate Bézier curve outlines, just like the chords in the bridge do.

Another interesting appearance of Bézier curves is in computer graphics. Whenever you use the pen tool, common in many image manipulation programs, you are drawing Bézier curves. More importantly, many computer fonts use Bézier curves to define how to draw their letters. Each letter is defined by up to several dozen control points

FIGURE 12. A pattern made from strings. [13].

and is drawn using a series of 3rd to 5th degree Bézier curves. This makes the letters scalable: They are clearly drawn and presented, no matter how much you zoom in on them.

This is the beauty of mathematics. It appears in places we would never expect and connects fields that appear entirely unrelated. Considering the fact that the curves were initially used for designing automobile parts, this is truly a display of the interdisciplinary nature of mathematics.

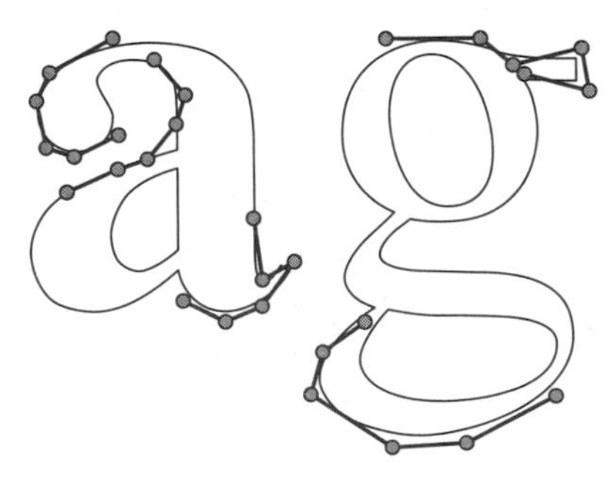

FIGURE 13. Some of the Bézier control points used to make "a" and "g" in the FreeSerif font (simplified).

Sources

[1] http://plus.maths.org/content/taxonomy/term/800
[2] http://plus.maths.org/content/category/tags/bezier-curve
[3] http://plus.maths.org/content/taxonomy/term/902
[4] http://plus.maths.org/content/taxonomy/term/674
[5] http://plus.maths.org/content/taxonomy/term/338
[6] http://plus.maths.org/content/taxonomy/term/21
[7] http://en.wikipedia.org/wiki/File:Jerusalem_Chords_Bridge.JPG
[8] http://plus.maths.org/content/finding-intersection-envelope
[9] http://plus.maths.org/content/changing-variables
[10] http://en.wikipedia.org/wiki/Pierre_Bézier
[11] http://plus.maths.org/content/formula-bezier-curve
[12] http://plus.maths.org/content/point-025025-lies-outline-curve
[13] http://www.stringartfun.com/product.php/7/free-boat-pattern
[14] http://sarcasticresonance.wordpress.com

Slicing a Cone for Art and Science

Daniel S. Silver

Albrecht Dürer (1471–1528), master painter and printmaker of the German Renaissance, never thought of himself as a mathematician. Yet he used geometry to uncover nature's hidden formulas for beauty. His efforts influenced renowned mathematicians, including Gerolamo Cardano and Niccolo Tartaglia, as well as famous scientists such as Galileo Galilei and Johannes Kepler.

We praise Leonardo da Vinci and other Renaissance figures for embracing art and science as a unity. But for artists such as Leonardo and Dürer, there was little science to embrace. Efforts to draw or paint directly from nature required an understanding of physiology and optics that was not found in the ancient writings of Galen or Aristotle. It was not just curiosity but also need that motivated Dürer and his fellow Renaissance artists to initiate scientific investigations.

Dürer's nature can seem contradictory. Although steadfastly religious, he sought answers in mathematics. He was outwardly modest but inwardly vain. He fretted about money and forgeries of his work, yet to others he appeared to be a simple man, ready to help fellow artists.

Concern for young artists motivated Dürer to write an ambitious handbook for all varieties of artists. It has the honor of being the first serious mathematics book written in the German language. Its title, *Underweysung der Messung,* might be translated as *A Manual of Measurement*. Walter Strauss, who translated Dürer's work into English, gave the volume a pithy and convenient moniker: the *Painter's Manual.*

Dürer begins his extraordinary manual with apologetic words, an inversion of the famous warning of Plato's Academy: Let no one untrained in geometry enter here:

FIGURE 1. A detail from Albrecht Dürer's *Melencolia I* from 1514 shows a magic square, in which each row, column, and main diagonal sum to the same total, in this case 34.

> The most sagacious of men, Euclid, has assembled the foundation of geometry. Those who understand him well can dispense with what follows here, because it is written for the young and for those who lack a devoted instructor.

The manual was organized into four books and printed in Nüremberg in 1525, just three years before the artist's death. It opens with the definition of a line, and it closes with a discussion of elaborate mechanical devices for accurate drawing in perspective. In between can be found descriptions of spirals, conchoids, and other exotic curves. Constructions of regular polygons are given. Cut-out models ("nets") of polyhedra are found. There is also an important section on typography, containing a modular construction of the Gothic alphabet. An artist who wishes to draw a bishop's crozier can learn how to do it with a compass and ruler. An architect who wants to erect a monument might find some sort of inspiration in Dürer's memorial to a drunkard, a humorous design complete with coffin, beer barrel, and oversized drinking mug.

Scholarly books of the day were generally written in Latin. Dürer wrote *Underweysung der Messung* in his native language because he wanted it to be accessible to all German readers, especially those with limited formal education. But there was another reason: Dürer's knowledge of Latin was rudimentary. Others later translated *Underweysung der Messung* into several different languages, including Latin.

There was no reason to expect that Dürer should have been fluent in Latin. As the son of a goldsmith, he was lucky to have gone to school at all. Fortunately for the world, Dürer displayed his unusual intelligence at an early age. "My father had especial pleasure in me, because he said that I was diligent in trying to learn," he recalled. He was sent to school, possibly the nearby St. Sebald parochial school, where he learned to read and write. He and his fellow students carried slates or wax writing tablets to class. (Johannes Gutenberg had invented a printing press only 40 years before, and books were still a luxury.) Learning was a slow, oral process.

When Dürer turned 13, he was plucked from school so that he could begin learning his father's trade. At that age, he produced a self-portrait that gives a hint of his emerging artistic skill. Self-portraits at the time were rare. Dürer produced at least 11 more during his lifetime.

What might have inspired a tradesman's son to study the newly rediscovered works of ancient Greek mathematicians such as Euclid and Apollonius? Part of the answer can be found in the intellectual atmosphere of Nüremberg at the time. In 1470, Anton Koberger founded the city's first printing house. One year later, he became Dürer's godfather. Science and technology were so appreciated in Nüremberg that the esteemed astronomer Johannes Müller von Königsberg, also known as Regiomontanus (1436–1476), settled there and built an observatory.

The rest of the answer can be found in the dedication of the *Painter's Manual:* "To my especially dear master and friend, Herr Wilbolden Pirckheymer, I, Albert Dürer wish health and happiness." This master, whose name is more commonly spelled Willibald Pirckheimer (1470–1530), was a scion of one of Nüremberg's most wealthy and powerful families. He was enormous in many ways, both physically and in personality, as well as boastful and argumentative. He was also a deeply knowledgeable humanist with a priceless library.

Pirckheimer's house was a gathering place for Nüremberg's brilliant minds. Despite the wide difference between their social rankings, Dürer and Pirckheimer became lifelong friends. Pirckheimer depended on Dürer to act as a purchasing agent during his travels, scouting for gems and other valuable items. Dürer depended on Pirckheimer for access to rare books and translation from Greek and Latin.

The word "Messung" meant more to Dürer than simple measurement. "Harmony" might have been closer to the mark. In his youth, possibly in 1494, Dürer had marveled over a geometrically based drawing of male and female figures by the Venetian artist Jacopo de' Barbari (about 1440–1516). Despite the fact that de' Barbari was unwilling to share his methods—or maybe because of it—Dürer became convinced that the secrets of beauty might be found by means of mathematics. Dürer was only 23 years old at the time. He devoted the remaining three decades of his life to the search, for as he reflected some years later, "I would rather have known what [de' Barbari's] opinions were than to have seen a new kingdom." Geometry, recovered from ancient works, lit his way.

The gravity of Dürer's quest can be sensed in his enigmatic engraving *Melencolia I,* shown in Figure 2. Now approaching the 500th anniversary of its creation, *Melencolia I* has been the subject of more academic debate than any other print in history. Is the winged figure dejected because she has tried but failed to discover beauty's secret? She holds in her hand an open compass. Above her head is a *magic square,* the first to be seen in Western art. (In a magic square, the numbers in each row and column, as well as the two main diagonals, add to the same total, in this case 34. In this one, the date of the engraving, 1514, appears in the lowest row.) Clearly Dürer's mathematical interests were not limited to geometry.

When he wrote the *Painter's Manual,* Dürer was approaching the end of a successful career. As a young man eager to learn more about the new science of perspective and to escape outbreaks of plague at home, he had made two trips to Italy. After the first journey, his productivity soared. Dürer's self-portrait of 1498 radiates an expanding confidence. (It was not the first time that plague encouraged scientific discovery, nor was it the last. In 1666 Isaac Newton escaped an outbreak of plague at Cambridge University, returning to his mother's farm, where he had the most profitable year that science has ever known.)

During his second visit to Italy, Dürer met with fellow artists, including the great master Giovanni Bellini, who praised his work. Dürer came to the conclusion that German artists could rise to the heights of the Italians, but only if they learned the foundations of their art. Such a foundation would prevent mistakes—and such a foundation required geometry. He returned with an edition of Euclid that bears his inscription: "I bought this book at Venice for one ducat in the year 1507—Albrecht Dürer."

Dürer purchased a house in Nüremberg and began to study mathematics. The *Painter's Manual* was not the book that he had originally planned to write. He had started work on *Vier Bücher von Menschlicher Proportion* ("Four Books on Human Proportion"), but soon realized that the mathematical demands that it placed on young readers were too great. The *Painter's Manual* was intended as a primer.

FIGURE 2. *Melencolia I* has been heavily debated among art historians. Is the angel's dejection caused by her inability to discover beauty's secret? The engraving reflects Dürer's mathematical interests. Dürer's mistaken belief that ellipses were egg-shaped is reflected in the shape of the bell opening. His quest to extend the mathematics behind beauty to artists led him to publish a primer that ended up influencing scientists as well as artists. Source: Wikimedia Commons.

FIGURE 3. Dürer produced at least a dozen self-portraits during his lifetime. The first, at age 13 *(top left)*, hinted at his emerging artistic gifts. A second, produced in 1498 at age 27 *(top right)*, showed him at the peak of a successful career, when his confidence was expanding and his productivity was soaring. A final self-portrait in 1522, at age 51 *(bottom)*, shows the artist after his body was ravaged by a disease that killed him a few years later. Source: Wikimedia Commons.

Work on the *Painter's Manual,* too, was temporarily halted when, in 1523, Dürer acquired 10 books from the library of Nüremberg mathematician Bernhard Walther (1430–1504). Walther had been a student of Regiomontanus and had acquired important books and papers from him. But Walther was a moody man who denied others access to this valuable cache. Walther died, but his library remained with his executors for two decades. Finally its contents had been released for sale. Dürer's precious purchases were chosen and appraised by Pirckheimer. It took Dürer two more years to absorb the ideas these books contained. The completion of the *Painter's Manual* would just have to wait.

It would be a book for artists, or so Dürer thought. Nevertheless he allowed himself to be carried aloft by mathematics. "How is it that two lines which meet at an acute angle which is made increasingly smaller will nevertheless never join together, even at infinity?" he asks (and proceeds to give a strange explanation). Later he writes: "If you wish to construct a square of the same area as a triangle with unequal sides, proceed as follows." It is difficult to imagine any artist of the 16th century making use of such ideas. These are the thoughts of a compulsive theoretician.

Time for Dürer to complete his *Painter's Manual* was running out. In December 1520, he had foolishly trekked to the swamps of Zeeland in the southwestern Netherlands, hoping to inspect a whale that had washed ashore. Alas, the whale had already washed away by the time he arrived. It was not a healthy place to visit, and the chronic illness that he contracted there eventually killed him after eight painful years.

Dürer's self-portrait of 1522 contrasts disturbingly with his earlier one. In the words of Strauss: "It represents Dürer himself in the nude, with thinned, disheveled hair and drooping shoulders, his body ravaged by his lingering disease." He fashioned himself as the Man of Sorrows.

No Matter How You Slice It

The subject [Conic Sections] is one of those which
seem worthy of study for their own sake.
—Apollonius of Perga

Although there is much in the *Painter's Manual* that rewards close examination, one specific area worthy of concentration is Dürer's treatment of *conic sections.* The techniques that Dürer found to draw them

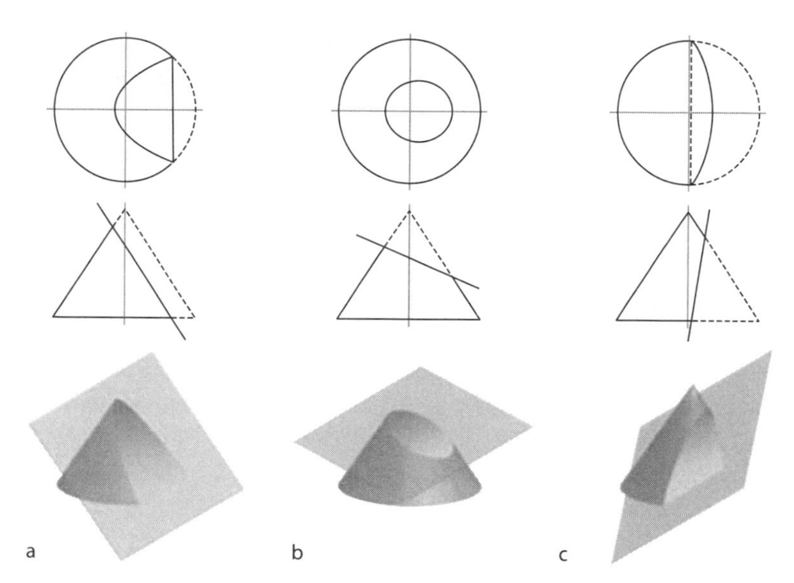

FIGURE 4. A plane slicing through a cone can produce several different shapes, called *conic sections*. Top and side views of the cone show how altering the angle of the plane results in a parabola *(a)*, ellipse, *(b)* or hyperbola *(c)*. Dürer's *Painter's Manual* aimed to show artists how to draw these shapes correctly.

anticipate the field of descriptive geometry that Gaspard Monge (1746–1818) developed later. The curves themselves would accompany a revolution in astronomy.

"The ancients have shown that one can cut a cone in three ways and arrive at three differently shaped sections," Dürer writes toward the end of Book I. "I want to teach you how to draw them."

Menaechmus (circa 350 B.C.), who knew Plato and tutored Alexander the Great, is thought to have discovered conic sections (often called simply "conics"). He found them while trying to solve the famous Delian problem of "doubling the cube." According to legend, terrified citizens of the Greek island of Delos were told by an oracle that plague would depart only after they had doubled the size of Apollo's cubical altar. Assuming that the altar had unit volume, the task of doubling it amounted to constructing a new edge of length precisely equal to the cube root of 2. Although the legend is doubtful, the Delian problem was certainly studied in Plato's Academy. Plato insisted on an exact solution accomplished using only ruler and compass.

Ingenious ruler-and-compass constructions abound in the *Painter's Manual.* Dürer's construction of a regular pentagon is particularly noteworthy. The construction method came not from Euclid but rather from one that had been taught by Ptolemy and is found in his *Almagest.* In 1837, the French mathematician Pierre Wantzel (1814–1848) proved that doubling the cube with ruler and compass is impossible. However, Menaechmus changed the rules of the game and managed to win. By intersecting a right-angled cone with a plane perpendicular to its side, he produced a curve that was later called a parabola. Then by intersecting two parabolas, chosen carefully, Menaechmus produced a line segment of length equal to the cube root of 2. (The parabola can be described by a simple equation $y^2 = 4px$. The positive number p is called the *latus rectum.*)

Menaechmus looked at other sorts of cones. When the cone's angle was either less than or greater than 90 degrees, two new types of curves resulted from their intersection with a plane. A century later, Apollonius of Perga (262–190 B.C.) called the three curves *parabola, ellipse,* and *hyperbola,* choosing Greek words meaning, respectively, comparison, fall short, and excess. Echoes are heard today in the English words such as parable, ellipsis, and hyperbole.

Today there is debate as to whether the terms originated with Apollonius. In any event, they were likely adapted from earlier terminology of Pythagoras (570–circa 495 B.C.) concerning a construction known as "application of areas." The interpretation in terms of angle is historically inaccurate but mathematically equivalent and simpler to state.

Apollonius' accomplishments went beyond nomenclature. He made a discovery that afforded a lovely simplification. Instead of using three different types of cones, as Menaechmus did, Apollonius used a single cone. Then by allowing the plane to slice the cone at different angles, he produced all three conics.

Associated with an ellipse or hyperbola are a pair of special points called *foci.* (For a circle, a special case of an ellipse, the distance between the two foci is zero.) Distances to the foci determine the curves in a simple way: The ellipse consists of those points such that the sum of distances to the foci is constant. Likewise, the hyperbola consists of points such that the difference is a constant.

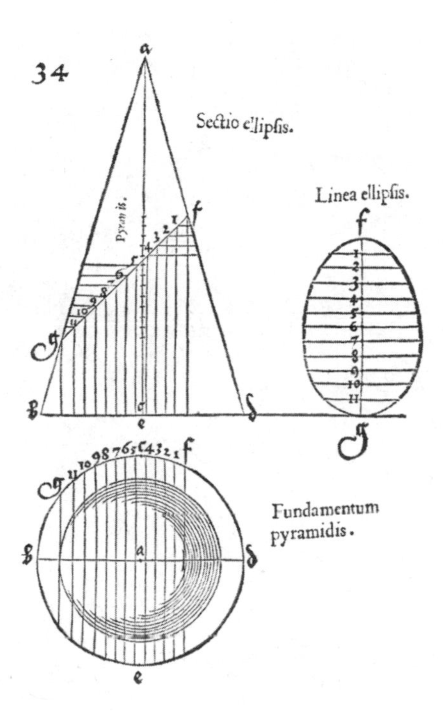

FIGURE 5. Dürer gave detailed instructions in his manual for how to transcribe the cutting of a cone with a plane *(seen in side and top view, left)* into an ellipse. He called ellipses "egg lines" because he believed, mistakenly, that they were wider on the bottom than on the top. (Unless otherwise indicated, all photographs are courtesy of the author.)

Tilt a glass of water toward you and observe the shape of the water's edge. It is an ellipse. So is the retinal image of a circle viewed from a generic vantage point.

Johannes Kepler (1571–1630) made the profound discovery that the orbit of Mars is an ellipse with the Sun at one focus. Kepler introduced the word "focus" into the mathematics lexicon in 1604. It is a Latin word meaning hearth or fireplace. What word could be more appropriate for the location of the Sun?

Kepler's letter to fellow astronomer David Fabricius (1564–1617), dated October 11, 1605, reveals that Kepler had read Dürer's description of conics:

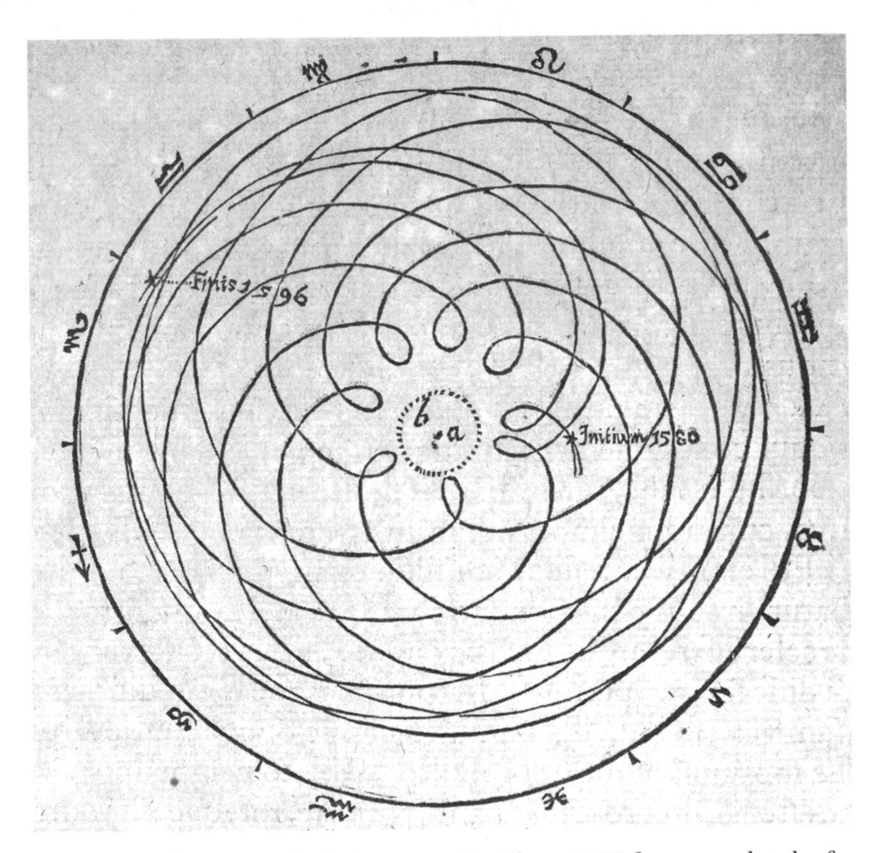

FIGURE 6. Johannes Kepler's *Astronomia Nova* from 1609 features a sketch of the retrograde motion of the planet Mars when viewed from Earth. Kepler made the discovery that the orbit of Mars is an ellipse with the Sun at one focus. His correspondence makes it clear that he had read Dürer's description of conics. Image courtesy of Linda Hall Library of Science, Engineering & Technology.

> So, Fabricius, I already have this: That the most true path of the planet [Mars] is an ellipse, which Dürer also calls an oval, or certainly so close to an ellipse that the difference is insensible.

In fact, Dürer used a more flavorful term for an ellipse, as we will see.

Nature's parabolas and hyperbolas are less apparent than the ellipse. A waterspout and the path of a cannonball have parabolic trajectories.

The wake generated by a boat can assume the form of a hyperbola, but establishing that fact requires more mathematics—or a boat.

Egg Lines

The significance of Dürer's treatment of conics is the technique that he used for drawing them, a fertile method of *parallel projection.* Art historian Erwin Panofsky observed that the technique was "familiar to every architect and carpenter but never before applied to the solution of a purely mathematical problem." In brief, Dürer viewed a cut cone from above as well as from the side, then projected downward. His trick was to superimpose the two views and then transfer appropriate measurements using dividers. In this way he relocated the curve from the cone to a two-dimensional sheet of paper.

Dürer's method was correct, but the master draftsman blundered while transferring measurements. He mistakenly believed that the ellipse was wider at the bottom of the cone than at the top, an understandable error considering the shape of the cone. As he transferred the distances with his divider, his erroneous intuition took hold of his hand.

Dürer writes, "The ellipse I call *eyer linie* [egg line] because it looks like an egg." Egg lines for ellipses can indeed be spotted in Dürer's work, such as in the bell in *Melencolia I.* Dürer knew no German equivalent of the Greek word "ellipse." The appellation he concocted drew attention to his error, and the egg line persisted in German works for nearly a century.

It is easy to understand why Kepler had an interest in Dürer's flawed analysis of the ellipse. For 10 years beginning in 1601, Kepler struggled to understand the orbit of Mars, a problem that had defeated Regiomontanus. Until he understood that the orbit was an ellipse, Kepler believed that it was some sort of oval. In fact, he specifically used the word "oval," a descendant of the Latin word "ovum" meaning egg.

Kepler was not the first to believe that a planet's orbit might be egg-shaped. Georg von Peuerbach (1423–1461), a teacher of Regiomontanus, had said as much in *Theoricae novae planetarum.* Published in Nüremberg in 1473 and reprinted 56 times, Peuerbach's treatise influenced both Copernicus and Kepler. The 1553 edition, published by Erasmus Reinhold (1511–1553), a pupil of Copernicus, included a comment about Mercury's orbit that might have caused Kepler to go back to the *Painter's Manual:*

> Mercury's [orbit] is egg-shaped, the big end lying toward his apogee, and the little end towards its perigee.

Later Kepler had this to say about the orbit of Mars:

> The planet's orbit is not a circle but [beginning at the aphelion] it curves inward little by little and then [returns] to the amplitude of a circle at [perihelion]. An orbit like this is called an oval.

Burning Mirrors

It is a safe bet that few artists in 16th century Germany felt the need to draw a parabola. In what seems like a marketing effort, Dürer tells his readers how it can be fashioned into a weapon of mass destruction.

The story that Archimedes set fire to an invading Roman fleet during the siege of Syracuse was well known in Dürer's time. Dioclese, a contemporary of Archimedes, explained the principle in his book *On Burning Mirrors,* preserved by Muslims in the 9th century.

Dioclese had observed something special about the parabola that had escaped the notice of Apollonius: On its axis of symmetry there is a point—a single focus—with the property that if a line parallel to the axis reflects from the parabola with the same angle with which it strikes, the reflected line passes through the focus. In physical terms, a mirror in the shape of a *paraboloid,* a parabola revolved about its axis, gathers all incoming light at the focus. Collect enough light, and whatever is at the focus will become hot.

Making an effective burning mirror is not a simple matter. Unless the parabola used is sufficiently wide, the mirror does not collect enough light. Dürer writes,

> If you plan to construct a burning mirror of paraboloid shape, the height of the cone you have to use should not exceed the diameter of the base—or this cone should be of the shape of an equilateral triangle.

Dürer goes on to explain why the angle of incidence of a light beam striking a mirror is equal to the angle of reflection. An elegant drawing of an artisan (possibly the artist) holding a pair of dividers does little to help matters (Figure 7). Dürer probably sensed that he was getting into a rough technical patch. He concludes the section desperately:

> The cause of this has been explained by mathematicians. Whoever wants to know it can look it up in their writings. But I have drawn my explanation . . . in the figure below.

Burning mirrors might have sounded useful to readers of the *Painter's Manual.* The first scientific evidence that Archimedes' mirrors might not have been such a hot idea had to wait for 12 years until René Descartes expressed doubts in his treatise *Dioptrique.* Nevertheless, since the time of Archimedes, burning mirrors, whatever their effectiveness, were constructed in a more practical, approximate fashion with sections of a sphere. (In 1668, Isaac Newton designed the first reflecting telescope on the principle of the burning mirror, with an eyepiece near the focus. He substituted a spherical mirror to simplify its construction.) It seems that Dürer liked parabolas and was determined to write about them.

Dürer invented German names for the parabola and the hyperbola as well as the ellipse. The parabola he called a *Brenn Linie* ("burn line"). "And the hyperbola I shall call *gabellinie* [fork line]," he writes, but he offers no explanation for his choice. Nor, it seems, has a reason been suggested by anyone else. Dürer might have been paying tribute to the many gabled houses of which Nüremberg was proud, the artist's own home near the Thiergärtnerthor included. In the *Painter's Manual,* Dürer constructs the hyperbola but has little to say about it.

Much of what Dürer knew about conic sections came from Johannes Werner (1468–1522). A former student of Regiomontanus, Werner was an accomplished instrument maker. He made contributions to geography, meteorology, and mathematics. A lunar impact crater named in his honor is not far from a crater named Regiomontanus.

Werner's *Libellus super viginti duobus elementis conicis* was published in 1522, at the time when Dürer was studying conics. The volume's 22 theorems were intended to introduce the author's work on the Delian problem. From handwritten notes, it appears that Werner died during his book's printing. (Werner's book soon became rare. It is reported that the Danish astronomer Tycho Brahe could not find a copy for sale anywhere in Germany.)

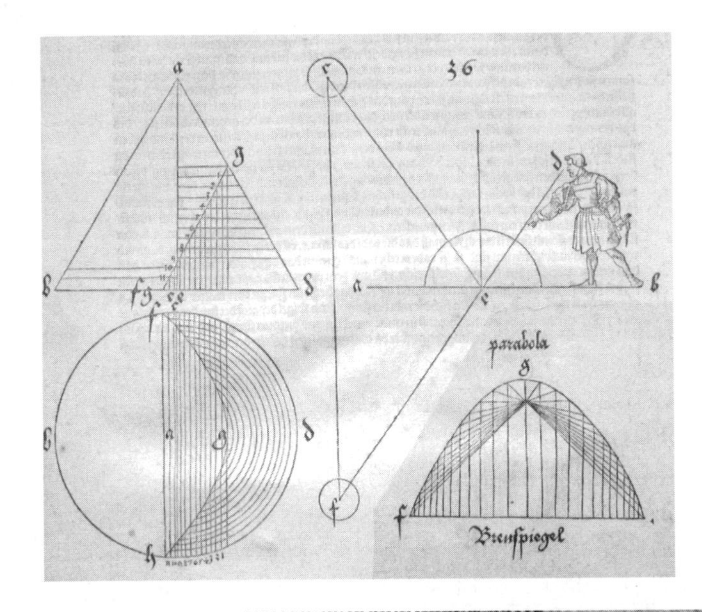

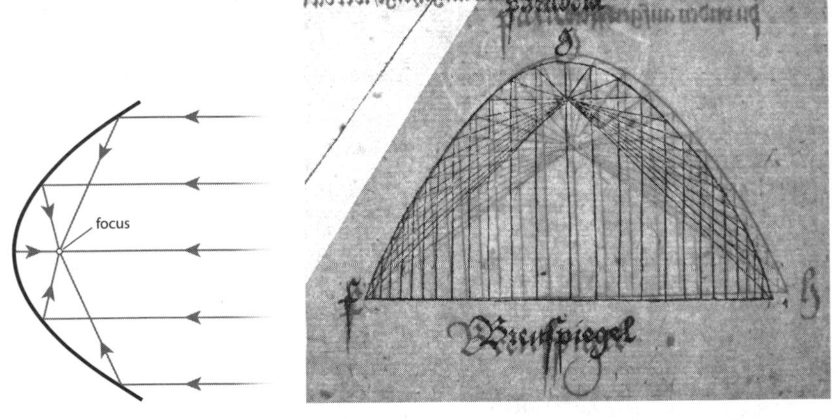

FIGURE 7. Dürer used a method called *parallel projection* when transcribing figures, such as this parabola, from three to two dimensions. He viewed a cut cone from above as well as from the side, then projected downward, superimposing the two views and then transferring appropriate measurements using dividers *(top)*. Dürer argued that parabolas correctly angled could be used as burning mirrors, heating what is at the focus *(bottom left)*. He tried to explain light angles with a drawing of an artisan *(top right)*. However, he miscalculated the placement of the focal point, so a correction had to be manually pasted into each copy of his 1525 publication. The original mistake is revealed behind the correction when the page is backlit *(bottom right)*.

Since 1508, Werner had been serving as priest at the Church of St. John, not far from Dürer's house. Like Pirckheimer, Werner acquired some of the rare books and papers that had been in Walther's possession. However, Werner knew no Greek and probably relied on Pirckheimer for translation. (His commentary on Ptolemy, published in 1514, is dedicated to Pirckheimer.) Like Dürer, Werner would have been a frequent visitor to Pirckheimer's house.

I believe that Dürer was inspired by Werner's novel construction of the parabola (Figure 8). The cone that he used was an oblique cone with vertex directly above a point on the base circle. A cut by a vertical plane produced the parabola. Regularly spaced, circular cross sections of the cone are in the lower diagram, each tangent to a point that lies directly below the vertex of the cone. The cutting plane is seen in profile as a line through points labeled *b* and *f*. By transferring the segments

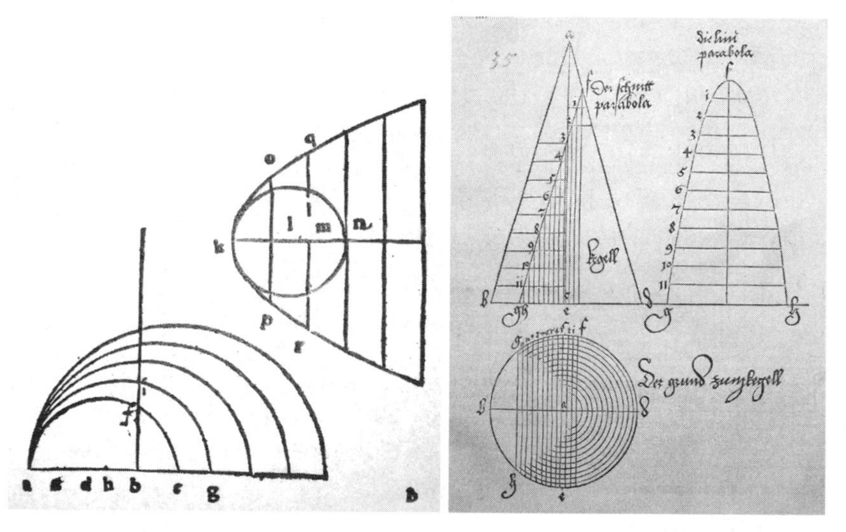

FIGURE 8. Johannes Werner, a contemporary of Dürer, made contributions to geography, meteorology, and mathematics. His parabola *(left)*, was published in 1522 at a time when Dürer was studying conics. It is likely that Dürer's parabola, published in 1525 *(right)*, was influenced by Werner's. The method of parallel projections that is often credited to Dürer might well have derived from Werner's construction. However, Werner used an oblique cone, whereas Dürer's was a right cone, so Werner's formula for the location of the focus no longer applied.

cut by the circles along the line, Werner produced the semi-arcs transverse to the line through *k* and *n* in the upper figure.

Had Dürer seen the picture, which is likely, Germany's master of perspective would have had no trouble imagining the tangent circles stacked in three dimensions, the smallest coming closest to his eye. The method of parallel projections that is often credited to Dürer might well have derived from Werner's construction.

In Appendix Duodecima of his book, Werner explains the reflective properties of the parabola to his audience. He also tells the reader how to locate the focus: Its distance from the vertex is one quarter of the length of the segment *ab*. (The length of *ab*, which is equal to the length of *kn*, is the *latus rectum* of the parabola—the distance between the slicing plane and the vertex of the cone.)

But Dürer used a right cone with vertex directly above the center of the circular base, so his cross-sectional circles became concentric rather than tangent to a single point, as in Werner's diagram.

Unfortunately for Dürer, Werner's formula for the location of the focus no longer applied. Whether Dürer computed the distance incorrectly or merely guessed, we do not know. However, in every copy of the 1525 publication a small piece of paper with the correct drawing had to be pasted by hand over the erroneous one. By holding the final product up to the light, Dürer's mistake is revealed.

Dürer and Creativity

For Albrecht Dürer, questions of technique eventually gave way to those of philosophy. In 1523, he wondered at the way "one man may sketch something with his pen on half a sheet of paper in one day . . . and it turns out to be better and more artistic than another's big work at which its author labors with the utmost diligence for a whole year."

The belief that divine genius borrows the body of a fortunate artist was common in Dürer's time. According to Panofsky, Leonardo da Vinci would have been perplexed had someone called him a genius. But Dürer had begun to see the creative process differently. For him, it became one of synthesis governed by trained intuition.

Dürer's last name was likely derived from the German word *tür*, meaning door. (His father was born in the Hungarian town of Ajtas, which is related to the Hungarian word for door, *ajtó*.)

FIGURE 9. Dürer engraving from 1514, titled *St. Jerome in His Study,* is noted for its use of unusual mathematical perspective, which invites the viewer into the snug chamber. It is rich with symbolism related to the theological and contemplative aspects of life in Dürer's time.

It is a fitting name for someone who opened a two-way passage between mathematics and art. As Panofsky observed, "While [the *Painter's Manual*] familiarized the coopers and cabinet-makers with Euclid and Ptolemy, it also familiarized the professional mathematicians with what may be called 'workshop geometry.'"

Dürer used geometry to search for beauty, but he never regarded mathematics as a substitute for aesthetic vision. It was a tool to help the artist avoid errors. However, the *Painter's Manual* demonstrates that mathematics and, in particular, geometry, meant much more to him. Four centuries after its publication, poet Edna St. Vincent Millay wrote, "Euclid alone has looked on Beauty bare." Dürer might have agreed.

Bibliography

Coolidge, Julian L. 1968. *A History of the Conic Sections and Quadric Surfaces.* New York: Dover Publications.

Dürer, Albrecht. 1525. *A Manual of Measurement [Underweysung der Messung].* Translated by Walter L. Strauss, 1977. Norwalk, Conn.: Abaris Books.

Eves, Howard. 1969. *An Introduction to the History of Mathematics, Third Edition.* Toronto: Holt, Rinehart and Winston.

Guppy, Henry, ed. 1902. *The Library Association Record, Volume IV.* London: The Library Association.

Heaton, Mrs. Charles. 1870. *The History of the Life of Albrecht Dürer of Nürnberg.* London: Macmillan and Co.

Herz-Fischler, Roger. 1990. "Durer's paradox or why an ellipse is not egg-shaped." *Mathematics Magazine* 63(2): 75–85.

Kepler, Johannes. 1937. *Johannes Kepler, Gesammelte Werke.* Vol. 15, letter 358, l. 390–392, p. 249. Walter von Dyck and Max Caspar, eds. Munich: C. H. Beck.

Knowles Middleton, William Edgar. 1961. "Archimedes, Kircher, Buffon, and the burning-mirrors." *Isis* 52(4): 533–543.

Koyré, Alexander. 2008. *The Astronomical Revolution.* Translated by R.E.W. Maddison. London: Routledge.

Pack, Stephen F. 1966. *Revelatory Geometry: The Treatises of Albrecht Dürer.* Master's thesis, School of Architecture, McGill University.

Panofsky, Erwin. 1955. *The Life and Art of Albrecht Dürer, Fourth Edition.* Princeton, N.J.: Princeton University Press.

Rupprich, Hans. 1972. "Wilibald Pirckheimer." In *Pre-Reformation Germany,* Gerald Strauss, ed. London: Harper and Row.

Russell, Francis. 1967. *The World of Dürer.* New York: Time Incorporated.

Strauss, Gerald, ed. 1972. *Pre-Reformation Germany.* London: Harper and Row.

Thausing, Moriz. 1882. *Albert Dürer: His Life and Works.* Translated by Fred A. Eaton, 2003. London: John Murray Publishers.

Toomer, Gerald J. 1976. "Diocles on burning mirrors." In *Sources in the History of Mathematics and the Physical Sciences 1.* New York: Springer.

Werner, Johannes. 1522. *Libellus super viginti duobus elementis conicis.* Vienna, Austria: Alantsee.

Westfall, Richard S. 1995. "The Galileo Project: Albrecht Dürer." http://galileo.rice.edu/Catalog/NewFiles/duerer.html.

Wörz, Adèle Lorraine. 2006. *The Visualization of Perspective Systems and Iconology in Dürer's Works.* Ph.D. dissertation, Department of Geography, Oregon State University.

High Fashion Meets Higher Mathematics

Kelly Delp

Try the following experiment. Get a tangerine and attempt to take the peel off in one piece. Lay the peel flat and see what you notice about the shape. Repeat several times. This can be done with many types of citrus fruit. Clementines work especially well.

Cornell mathematics professor William P. Thurston used this experiment to help students understand the geometry of surfaces. Thurston, who won the Fields Medal in 1982, was well known for his geometric insight. In the early 1980s, he made a conjecture, called the *geometrization conjecture*, about the possible geometries for three-dimensional manifolds. Informally, an n-dimensional manifold is a space that locally looks like $\mathbb{R}^n$.

Although Thurston proved the conjecture for large classes of three-manifolds, the general case remained one of the most important outstanding problems in geometry and topology for 20 years. In 2003 Grigori Perelman proved the conjecture. The geometrization conjecture implies the Poincaré conjecture, so with his solution Perelman became the first to solve one of the famed Clay Millennium Problems. (The November 2009 issue of *Math Horizons* ran a feature on Perelman.)

The story of Thurston's geometrization conjecture and the resolution of the Poincaré conjecture drew attention from reporters and other writers outside of the mathematical community. One person who happened upon an account of Thurston and his work was the creative director of House of Issey Miyake, fashion designer Dai Fujiwara. In a letter to Thurston, Fujiwara described how he felt a connection with the geometer, as he had used the same technique of peeling fruit to explain clothing design to students new to the subject. Designers also practice the art of shaping surfaces from two-dimensional pieces.

FIGURE 1. Dai Fujiwara and William P. Thurston, Paris Fashion Week, spring 2010. Photo by Nick Wilson.

Fujiwara felt that Thurston's three-dimensional geometries could provide a theme for Issey Miyake's ready-to-wear fashion line. Thurston, who in 1991 had organized (along with his mother, Margaret Thurston) what was perhaps the first mathematical sewing class as part of the Geometry and Imagination Workshop, agreed that there was potential for connection. Thus the collaboration was born. The Issey Miyake collection inspired by Thurston's eight geometries debuted on the runway at Paris Fashion Week in spring 2010.

Geometry of Surfaces

Before discussing the fashion show and the geometry of three-manifolds, we discuss geometries of two-dimensional objects, or surfaces. Examples of surfaces include the sphere, the torus, and the Möbius band. Any (orientable) surface can be embedded in $\mathbb{R}^3$, and this fact

Beauty Is Truth, Truth Beauty—That Is All Ye Know on Earth, and All Ye Need to Know

This famous and provocative quotation of John Keats is echoed on the emblem of the Institute for Advanced Study, where I took my first job after graduate school. After reading an account of my mathematical discoveries concerning eight geometries that shape all three-dimensional topology, Dai Fujiwara made the leap to write to me, saying that he felt in his bones that my insights could give inspiration to his design team at Issey Miyake. He observed that we are both trying to understand the best three-dimensional forms of two-dimensional surfaces, and he noted that we each, independently, had come around to asking our students to peel oranges to explore these relationships. This notion resonated strongly with me, for I have long been fascinated (from a distance) by the art of clothing design and its connections to mathematics.

Many people think of mathematics as austere and self-contained. To the contrary, mathematics is a very rich and very human subject, an art that enables us to see and understand deep interconnections in the world. The best mathematics uses the whole mind, embraces human sensibility, and is not at all limited to the small portion of our brains that calculates and manipulates symbols. Through pursuing beauty we find truth, and where we find truth we discover incredible beauty.

The roots of creativity tap deep within to a place we all share, and I was thrilled that Dai Fujiwara recognized the deep commonality underlying his efforts and mine. Despite literally and figuratively training and working on opposite ends of the Earth, we had a wonderful exchange of ideas when he visited me at Cornell. I feel both humbled and honored that he has taken up the challenge to create beautiful clothing inspired by the beautiful theory that is dear to my heart.

—William P. Thurston

allows us to measure distances on the surface. Let's think about the sphere with radius 1 centered at the origin. This sphere is described by the familiar equation $x^2 + y^2 + z^2 = 1$. Choose any two points on the sphere, say the north pole (0, 0, 1) and the south pole (0, 0, –1).

There are two natural ways to define the distance between these points. The first way is to assign the distance between points to be the usual Euclidean distance in $\mathbb{R}^3$. In this metric, the distance between the poles is 2, the length of the diameter.

For another metric, assign the distance to be the minimum length of any path on the sphere that starts at one pole and ends at the other; this distance would be the length of the shortest arc of a great circle between them. Now the distance between the poles would be π. This latter metric is more appropriate; after all, when traveling from Buffalo to Sydney, the best way is not by drilling through the center of the Earth.

Two manifolds are *topologically equivalent* if there is a continuous bijection, with a continuous inverse, between them. The bijection is called a *homeomorphism*, and we say that the manifolds are *homeomorphic*. Under this equivalence relation, all of the surfaces in Figure 2 are spheres.

Each of these spheres can be equipped with a metric from $\mathbb{R}^3$, as previously described, by measuring the shortest path in the surface. Even though they are topologically equivalent, as metric spaces they are very different. Two surfaces are *metrically equivalent* if there is a distance-preserving map, called an *isometry*, between them. One quantity that is preserved under isometries is Gaussian curvature. Recall that the Gaussian curvature is a function k from a surface S to the real numbers, where $k(p)$ is the product of the principal curvatures at p; roughly speaking, $k(p)$ gives a measure of the amount and type of bending of the surface at a point p. At a point of positive curvature, all of the (locally length minimizing) curves through p bend in the same direction; in

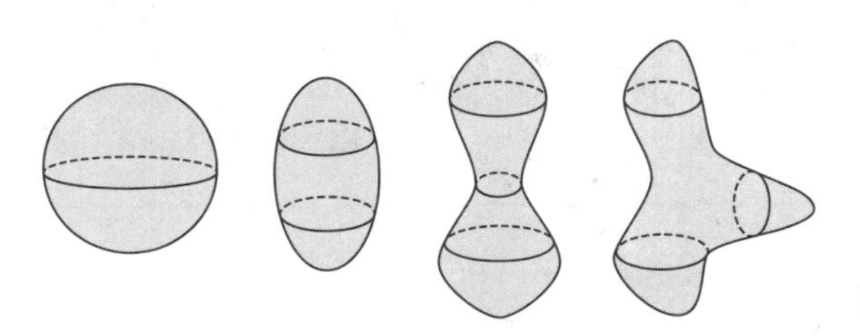

FIGURE 2. Spheres.

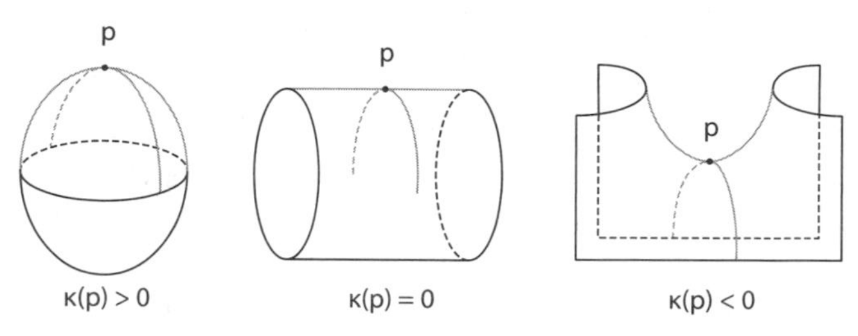

FIGURE 3. Gaussian curvature.

negative curvature, the surface has curves that bend in opposite directions. A surface that contains a straight line through a point p gives an example of zero curvature (Figure 3).

You should be able to identify points of negative curvature in the right two spheres in Figure 2. The second sphere has only positive curvature, though the curvature is greater at the north pole than at the equator. The first sphere is the most symmetric and has constant curvature for $k(p) = 1$ for all p.

The round sphere is one of three model two-dimensional geometries. The other two model geometries are the Euclidean plane, which has constant zero curvature, and the hyperbolic plane, which has constant curvature of -1. Every (compact and smooth) surface supports a metric of exactly one type of constant curvature: positive, negative, or zero. Although the sphere has many different metrics, it cannot have a Euclidean or hyperbolic metric. We give examples of surfaces of the latter two types.

Euclidean surfaces, such as a torus, can be constructed from pieces of the Euclidean plane. We can start with a rectangle in the Euclidean plane, a sheet of paper works nicely as a model, and tape together opposite sides of the piece of paper (mathematically, this is done by

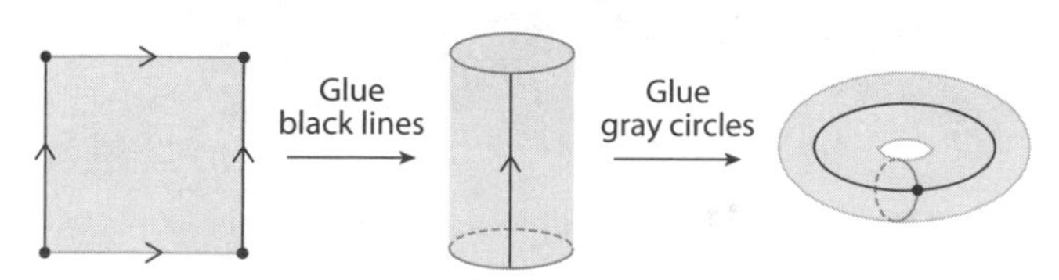

FIGURE 4. Building a torus.

creating a quotient space). If we stop at this point, we have a Euclidean cylinder. If we identify the opposite two boundary circles, we create a torus with a Euclidean metric inherited from the original rectangle. Of course, if you try to do this with your paper cylinder, you will find it impossible. You can come close by folding and creasing the cylinder, but the final object does not look very much like a torus. We do not allow creasing as a legal construction technique, as the corresponding mathematical object would not have a tangent sphere along the crease. If we had a fourth dimension to bend into, we could tape together the opposite circles without distorting the metric.

Finally, we describe an example of a hyperbolic surface. Figure 5a pictures a crocheted model of a piece of the hyperbolic plane, conceived and constructed by Daina Taimina. We point out one difference between the hyperbolic plane and the Euclidean plane, which can be seen in this model. If one crocheted a Euclidean disc, the number of stitches in concentric circles would increase linearly; this phenomenon occurs because the circumference of a circle in the Euclidean plane is a linear function of the radius: $2\pi r$. In the hyperbolic plane, the circumference of the circle grows *exponentially* with the radius—$2\pi\sinh(r)$—creating the wavy surface seen in Figure 5a.

Figure 5b shows a regular octagon with 45-degree interior angles in the hyperbolic plane. Note that such a polygon could not occur in the Euclidean plane, where regular octagons have interior angles of 135 degrees. If we identify every other side of this octagon, we create a *hyperbolic pair of pants*, shown in Figure 5c. Note that "pair of pants" is the name geometric topologists use to refer to surfaces of this homeomorphism type, which are important building blocks for all hyperbolic surfaces.

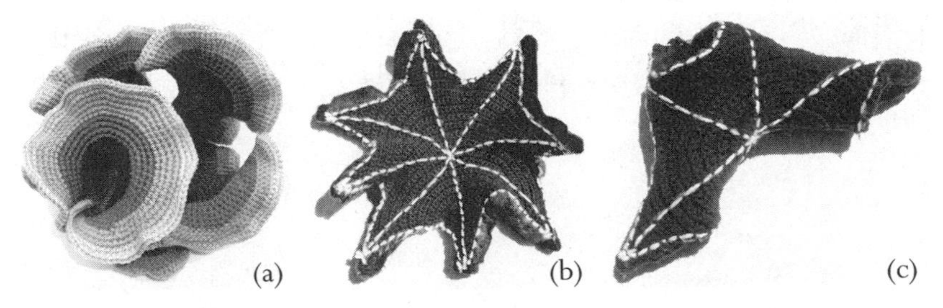

FIGURE 5. Daina Taimina's hyperbolic crocheting: (a) hyperbolic plane, (b) hyperbolic octagon, and (c) hyperbolic pair of pants. © 2009. From *Crocheting Adventures with Hyperbolic Planes* by Daina Taimina. Reproduced by permission of Taylor and Francis Group, LLC, a division of Informa plc and Daina Taimina.

Orbifolds, Three-Dimensional Geometries, and Design

Essentially, Thurston's geometrization conjecture states that any three-manifold can be decomposed into finitely many pieces, each of which supports a metric modeled on one of eight geometries: the three-dimensional analogue of spherical, hyperbolic, or Euclidean space, or one of five other possible geometries: $S^2 \times \mathbb{R}$, $H^2 \times \mathbb{R}$, *Nil*, *Solv*, or *universal cover of SL* (2,$\mathbb{R}$). If a closed manifold supports one of these eight geometries, it cannot support a metric of any of the other seven types.

An example of a Euclidean three-manifold is the three-dimensional torus. To mathematically construct this manifold, start with a solid Euclidean cube, which can be described as the set of points (x, y, z) in $\mathbb{R}^3$ such that $0 \leq x, y, z \leq 1$. Identify opposite faces of the cube by distance-preserving maps. More explicitly, identify the top face $(z = 1)$ to the bottom face $(z = 0)$, by defining an equivalence relation on the cube that identifies $(x, y, 0)$ with $(x, y, 1)$. Analogous identifications can be made between the front and back faces, and with the left and right faces.

The room you are sitting in, if it is roughly cubical, provides a good model for this space. Once the floor and the ceiling have been identified, when you look straight up you will see the bottom of your feet. If you would like to experience different geometries of three-dimensional manifolds, try Jeff Week's program *Curved Spaces*, available at his website: http://geometrygames.org/.

Using the eight three-dimensional geometries as inspiration for a fashion line seemed like a difficult endeavor as clothing is essentially two-dimensional. To learn about these geometries, and to exchange ideas, Fujiwara visited Thurston at Cornell. After returning to Japan, he continued to exchange ideas with Thurston, and the topologist Kazushi Ahara from Meiji University gave a series of lectures about the geometries to the design team. The designers were a somewhat apprehensive audience, and Ahara promised not to use certain words, such as "equation" or "trigonometric function" during his lectures.

So what was the result of the collaboration? How did Fujiwara accomplish the difficult task of representing the eight geometries in the Issey Miyake line of fashion? At the time of writing this article, several videos from the fashion show can be found on YouTube, so the interested reader can form his or her own opinion. The videos can be found by searching for "Issey Miyake Fashion Show: Women's Ready to Wear Autumn/Winter 2010."

In an interview, Fujiwara explained how his collection was "an expression of space." From this statement, one gets the impression that the designers mainly used the mathematics as inspiration for their work, rather than creating an explicit illustration of the geometries. There was also a more concrete, somewhat poetic connection between the collection and the geometries. Before describing this connection, we need to introduce one more mathematical object: the *orbifold*.

An orbifold is a manifold with singularities. Like manifolds, we can equip orbifolds with metrics modeled on a specific geometry. For our purposes, it should be sufficient to understand two related examples. Let's start with the two-dimensional case. A cone, with cone angle of $2\pi/3$, is an example of a Euclidean orbifold with one singular point of order, 3. This cone can be constructed in two ways. In a method similar to the construction of the torus, we can cut a wedge from the circle with angle $2\pi/3$ and tape up the sides. We could also cut just one slit from the edge of the circle to the center, and then roll up the disk so that it wraps around itself three times. Mathematically, this process can be described as taking the quotient of the disk by a rotation. Away from the cone point, every point has a small neighborhood so that the metric looks just like a small disk in $\mathbb{R}^2$.

The higher dimensional analog of the cone can be constructed from a solid cylinder. Again, we can think about the construction in two ways: either as cutting a wedge and gluing opposite sides, or by this process of rolling up the cylinder so that it wraps around itself three times. We see that in three-dimensional spaces our singular sets can be one-dimensional. We have a whole line segment of singularities labeled with a 3.

The important point is that a particular three-dimensional orbifold can belong to at most one of the eight geometric classes and that singular sets can be one-dimensional. Sometimes these singular sets have

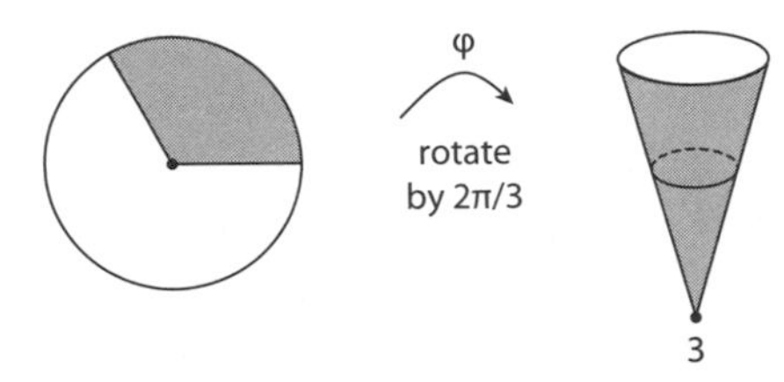

FIGURE 6. Euclidian cone.

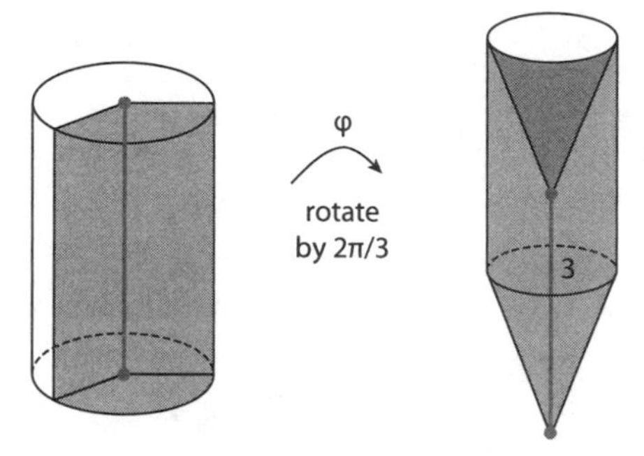

FIGURE 7. Three-dimensional Euclidian orbifold.

several components, which are linked together. The three-sphere S^3, which is the set of points distance 1 from the origin in $\mathbb{R}^4$, is a three-dimensional manifold and is the model space for one of the eight geometries. However, for the *orbifold* S^3, with a one-dimensional singular set, the metric class depends on how the singular set is sitting inside S^3. A table of links is shown in Figure 8. Each link corresponds to the orbifold S^3 with the given link as a singular set of order 2. Each of the orbifolds carries a different one of the eight geometries.

Thurston drew the links in Figure 8. They were one of the many ideas that he shared with Fujiwara. The links intrigued Fujiwara, and

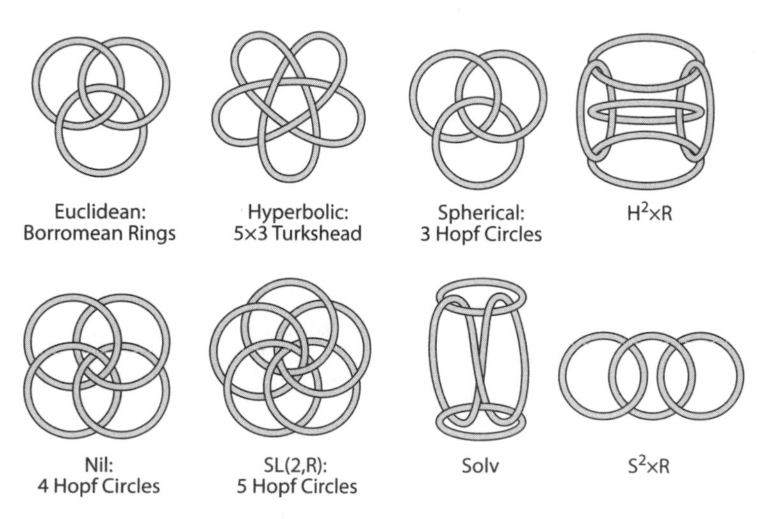

FIGURE 8. Illustrations of orbifold representatives of the eight geometries.

Figure 9. Issey Miyake's spring 2010 collection on the runway at Paris Fashion Week. Courtesy of Frederique Dumoulin/Issey Miyake.

they appeared as an integral part of several of the pieces in the fashion line, as seen on the models on the runway.

In an article written for the fashion magazine *Idoménée*, Thurston gave the following comment about the collection:

> The design team took these drawings as their starting theme and developed from there with their own vision and imagination. Of course it would have been foolish to attempt to literally illustrate the mathematical theory—in this setting, it's neither possible nor desirable. What they attempted was to capture the underlying spirit and beauty. All I can say is that it resonated with me.

Acknowledgment

I am but one of many who were influenced by Bill Thurston and saddened by his death on August 21, 2012. I am grateful for the time we spent playing with mathematics.

The Jordan Curve Theorem Is Nontrivial

Fiona Ross and William T. Ross

1. Introduction

The classical Jordan curve theorem (JCT) says,

> Every Jordan curve (a non-self-intersecting continuous loop in the plane) separates the plane into exactly two components.

It is often mentioned just in passing in courses ranging from liberal arts mathematics courses, where it is an illuminating example of an "obvious" statement that is difficult to prove, to undergraduate and graduate topology and complex analysis, where it tends to break the flow. In complex analysis, it is especially given short shrift. There are several reasons for this short shrift. For one, a professor has bigger fish to fry. There are the theorems of Cauchy, Hadamard, Morera, and the like, which comprise the nuts and bolts of complex analysis, and so the Jordan curve theorem appears to be a mere curiosity. Second, there is insufficient time to even outline a proof that an uninitiated student can really appreciate. Third, the result is clearly "obvious," and professors do not want to put themselves, or their students, through a complicated proof of a theorem which seems to need no proof at all.

A complex analysis teacher who recognizes the need to at the least mention the JCT might hastily draw a circle or an ellipse on the board to point out the interior and exterior regions—knowing full well that the students are not impressed. The slightly more ambitious teacher might make some weak attempt (without practicing this before class) at drawing something more complicated and invite the students to identify the interior and exterior regions. The students always do so quickly and are still not impressed and don't see the real difficulties. Maybe the teacher gives a slightly more formal "proof" of the JCT:

> Start from a point not on the curve and draw a straight line from that point to the outside of the whole drawing. If the line meets the curve an odd number of times, you are on the interior. If the line meets an even number of times, you are on the exterior.

For most curves a (nonartist) teacher might draw in class, this "counting crossings" method is relatively easy, and students still don't see the real difficulties or why this is not really a valid proof of the JCT. The Jordan curve theorem lesson usually ends with a few mumbled words like, "Well, trust me on this one. Things can get pretty complicated out there, and to make this all exact takes a lot of work. So, let's move on to . . ."

We certainly understand the issues mentioned above in teaching the Jordan curve theorem and the need to move on to more relevant topics. We write this note for those who wish to appreciate both the artistic exploration and the history behind the JCT. Thus we unite two lines of discourse on the nature of nontriviality. In our opinion, the JCT is a wonderful result because it exposes us to amazing, pathological, counterintuitive examples, such as nowhere differentiable curves or curves with positive area (see below). So, in a way, not appreciating the JCT is driven by a lack of imagination, in thinking that Jordan curves are nothing more than circles or a couple of wavy lines a teacher hastily draws on a blackboard.

The JCT not only inspires mathematicians to dream up these fantastical examples, which almost mock the technical definition of a curve, but it also inspires artists. The open curve artwork of Mø Morales (see http://virtualmo.com) and the maze art of Berg [l] are well known and much admired. Kaplan [10] and Pedersen and Singh [12] have developed algorithms for computer-generated mazes and labyrinths. Perhaps most relevant are the computer-generated Jordan curve artwork of Bosch [2] and the TSP art of Bosch and Kaplan [3]. Before becoming familiar with these artists' work, the first author (Fiona) made the drawings described here to show that Jordan curves are not the cold, abstract, boring objects we might think they are. Instead, they can tell a story. They can help mathematicians make a better case to their students, and to themselves, that there is something nontrivial going on here—indeed, something beautiful. We see through some hand-drawn artwork that the interior and exterior regions guaranteed by the JCT are not so easy to identify, and when they are identified, they can be just wonderful to look at.

Just as curves can be used to tell a story, the proof of the JCT is a great story by itself—from the Bohemian monk who had to convince us that the JCT needed a proof, to the Frenchman who gave us an unconvincing one, to the mathematicians who opened up the Pandora's box of terrifying pathological examples of curves, to the final proof, to Jordan's redemption. The artwork inspired by the JCT, and how it can inspire mathematicians and teachers, as well as the history of the result, are illuminated here.

2. The Artwork

The black-and-white hand drawings that are Jordan curves drawn by the first author (Figures 1 and 2), were created to give the reader an opportunity to explore the idea of a curve as leading the viewer through a story and also to invite the reader to apply the "proof" of the JCT mentioned earlier (counting the number of crossings needed to exit the drawing) to determine which parts of the figures lie in the interior or exterior. As a visual artist, the first author has always found inspiration in the way curves appear in the structure of the natural world. The curves that appear in geology and biology, such as anafractuous (twisting and turning), concentra, labyrinthine, phyllotaxy (dynamic spirals), vermiculate (wormlike arrangements), and ripple and dune formations have provided a starting point for many of her artworks. In exploring the physical and metaphorical structure of unicursal labyrinths, the first author created a series of self-portraits in landscapes rendered from a single line. The most famous of unicursal, or single path, labyrinths features prominently in the mythological tale of Theseus and the Minotaur. Labyrinth forms are also used as guides to meditation and prayer, giving the person who travels the labyrinth a sense of their very selves as the most complex of labyrinths.

The authors began a discussion of the JCT after the second author (William) saw one of the first author's unicursal labyrinths (*Blue Introverted Unicursal Labyrinth*, Figures 3 and 4). In this drawing, the first

FIGURE 1. Fiona Ross, *When We Could Be Diving for Pearls*, 9¾″ × 6″, 2011, Micron ink on Denril paper.

author made a point of not uniting the curve that formed the labyrinth. Just as there is no metaphorical separation between the journey and the traveler, there was, deliberately, no separation of the interior or exterior of the labyrinth on the paper. Intrigued by the conversations with the first author, who is a mathematician, and by Felix Klein's statement,

> Everyone knows what a curve is, until he has studied enough mathematics to become confused through the countless number of possible exceptions,

the first author combined the idea of a curve leading the viewer through a meaningful story together with the mathematical notion of the

FIGURE 2. Fiona Ross, *A Thread in the Labyrinth*, 6″ × 6″, 2011, Micron ink on Denril paper.

complexities encountered when trying to prove the JCT. She began a new series of drawings (*When we could be diving for pearls* (Figure 1) and *A thread in the labyrinth* (Figure 2) rendered from a single (nonintersecting) curve without beginning or end, i.e., a Jordan curve. In these new drawings, the unicursal labyrinth is no longer open, as in the earlier *Blue Introverted Unicursal Labyrinth* (Figures 3 and 4), but closed. The interior and exterior spaces in the new drawings are difficult to identify. Making it difficult to identify inside and outside spaces gives the "closed" drawings in Figures 1 and 2 a different meaning than the "open" drawings of Figures 3 and 4. The Jordan curve drawings have a more secretive content. They seem to hold their figurative breath as the inked fingers of the external space unsuccessfully reach in and touch the interior, whereas the non-Jordan curve drawings invite entry into the most remote areas of the work. The first author finds both the forms of content (open and closed) interesting to work with and is grateful to be given the opportunity to point out where art and mathematics

FIGURE 3. Fiona Ross, *Blue Introverted Unicursal Labyrinth*, 14″ × 8″, 2010, Ink on Yupo paper.

intersect. The Jordan curve drawings specifically created for this article (Figures 1 and 2) were hand drawn with Micron ink markers or graphite on Denril drafting film.

3. A Historical Perspective of the Jordan Curve Theorem

Just as the JCT artwork in the previous section shows that curves can tell a tale, the JCT itself is an interesting story. The JCT says that the plane is separated into two components by a Jordan curve.

We traditionally call the bounded component the "interior" and the unbounded component the "exterior." Bernard Bolzano [8,13] realized that the problem was nontrivial and officially posed it as a theorem needing a proof. Here is Bolzano's "prophetic version of the celebrated Jordan curve theorem" (8, p. 285):

FIGURE 4. *Blue Introverted Unicursal Labyrinth*, detail.

> If a closed line lies in a plane and if by means of a connected line one joins a point of the plane which is enclosed within the closed line with a point which is not enclosed within it, then the connected line must cut the closed line.

Bolzano also realized that the current notions of curve at the time were in desperate need of proper definitions! The first proof, and hence the name Jordan curve theorem, was given in 1887 by Camille Jordan in his book *Cours d'analyse de l'École Polytechnique* [9] but was regarded by many to be incorrect. As an interesting side note, Thomas Hales [7] believes that the charges of incorrectness against Jordan's original proof were trumped up and that the errors are merely aesthetic. With a few relatively minor changes, he claims to make Jordan's proof rigorous.

If the Jordan curve is a polygon, one can prove the JCT using the "counting crossings" method mentioned earlier. This proof, once cleaned up a bit to make it completely rigorous, is the standard way to prove the JCT for polygons.

When the curve is not a polygon, all sorts of pathologies can occur. For the student, who may be reminded of defining the interior of the curve by using the rule learned from calculus: "When you travel around the curve, the interior is on your left," it is important to notice that one is tacitly assuming that there is a continuously turning normal to the curve (your left hand). There are examples of functions f: $[0, 1] \to \mathbb{R}$ that are continuous everywhere but differentiable nowhere, i.e., Weierstrass' famous example [5]

$$f(x) = \sum_{n=0}^{\infty} a^n \cos(b^n \pi x),$$

where $0 < a < 1$, $b > 1$ is an odd integer, and $ab > 1 + (3\pi/2)$ (Figure 5). The graph of f is jagged everywhere and can be used to define a Jordan curve that is also jagged everywhere. Such a curve does not have a well-defined tangent (and is hence normal). These types of jagged curves also serve to point out how our informal proof of the JCT

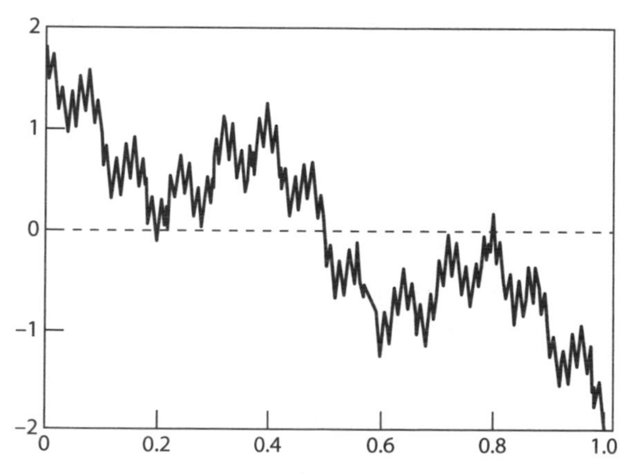

FIGURE 5. A Weierstrass nowhere differentiable function.

for polygonal type regions (count the number of times you cross the curve) breaks down since you might cross the curve an infinite number of times when exiting the drawing, making the crossing parity (i.e., odd or even) undefined.

Even more pathological is an example of Osgood [11] of a Jordan curve with positive area. Jordan showed that if the curve is rectifiable (of finite length), then it has zero area. These positive-area Osgood curves are quite wild and fascinating, and the reader is directed to a wonderful Mathematica Demonstration by Robert Dickau [4], where one constructs a Jordan curve with any desired (finite) area.

Otto Veblen [15] finally gave what many regard as the first correct proof of the Jordan curve theorem, and others have followed with different proofs, including ones using formal logic [6].

4. Final Thought

In this article, we have seen that a Jordan curve is much more than a circle or ellipse. It can be jagged at every point or perhaps impossible to visually determine the interior from the exterior or perhaps even of positive area, and, as seen through the artwork in this article, can lead the viewer through a story. In fact, the Jordan Curve Theorem is a wonderful story in itself—a seemingly obvious result, which Bolzano had to convince people was not and whose original proof was not convincing and which inspired mathematicians to create pathological, counterintuitive, and fascinating examples. It is then fitting that Jordan curves inspire art and, more importantly, help mathematicians make a better case to the students that the JCT is non-trivial both as a mathematical result and as a work of art.

References

[1] C. Berg, *Amazing Art: Wonders of the Ancient World.* Harper Collins, New York, 2001.

[2] R. Bosch, *Simple-closed-curve sculptures of knots and links,* J. Math. Arts 4 (2010), pp. 57–71. http://www.tandfonline.com/doi/abs/10.1080/17513470903459575#.UYQMp0pTn0U.

[3] R. Bosch and C. Kaplan, *TSP Art,* Bridges Conference Proceedings 2005, 303–310, http://www.cgl.uwaterloo.ca/~csk/projects/tsp/.

[4] R. Dickau, Knopp's Osgood Curve Construction, Wolfram Demonstrations Project, http://demonstrations.wolfram.com.KnoppsOsgoodCurveConstruction (user needs the *Mathematica* player installed to see this demonstration.

[5] B. R. Gelbaum and M. H. J. Olmsted, *Counterexamples in analysis,* Corrected reprint of the second (1965), Dover Publications Inc., Mineola, N.Y., 2003.

[6] T. C. Hales, *The Jordan curve theorem, formally and informally*, Amer. Math. Monthly 114 (2007), pp. 882–894.

[7] T. C. Hales, *Jordan's proof of the Jordan curve theorem*, Studies in Logic, Grammar and Rhetoric 10 (2007), pp.45–60.

[8] D. M. Johnson, *Prelude to dimension theory: the geometrical investigations of Bernard Bolzano,* Arch. History Exact Sci. 17 (1977), pp. 262–295.

[9] C. Jordan, *Cours d'analyse de l'École Polytechnique*, Gauthier-Villars, Paris, 1887.

[10] C. Kaplan, *Vortex maze construction*, J. Math. Arts 1 (2007), pp. 7–20.

[11] W. F. Osgood, *A Jordan curve of positive area*. Trans. Amer. Math. Soc. 4(1) (1903), pp. 107–112.

[12] H. Pedersen and K. Singh, *Organic labyrinths and mazes,* NPAR Conference Proceedings, (2006), 79–86, http://dx.doi.org/10.1145/1124728.1124742.

[13] B. B. Van Rootselaar, *Dictionary of Scientific Biography;* Vol. II, Charles Scribners Sons, New York, 1970, pp.273–279.

[14] H. Sagan, *A geometrization of Lebesgue's space-filling curve,* Math. Intelligencer 15 (1993), pp. 37–43.

[15] O. Veblen, *Theory on plane curves in non-metrical analysis situs*, Trans. Amer. Math. Soc. 6 (1905), pp. 83–98.

Why Mathematics? What Mathematics?

Anna Sfard

"Why do I have to learn mathematics? What do I need it for?" When I was a school student, it never occurred to me to ask these questions, nor do I remember hearing it from any of my classmates. "Why do I need history?"—yes. "Why Latin?" (yes, as a high school student I was supposed to study this ancient language)—certainly. But not, "Why mathematics?" The need to deal with numbers, geometric figures, and functions was beyond doubt, and mathematics was unassailable.

Things changed. Today, every other student seems to ask why we need mathematics. Over the years, the quiet certainty of the mathematics learner has disappeared: No longer do young people take it for granted that everybody has to learn math, or at least the particular mathematics curriculum that is practiced with only marginal variations all over the world. The questions, "Why mathematics? Why *so much* of it? Why 'for all'?" are now being asked by almost anybody invested, or just interested, in the business of education. Almost, but not all. Whereas the question seems to be bothering students, parents, and, more generally, all the "ordinary people" concerned about the current standards of good education, the doubt does not seem to cross the minds of those who should probably be the first to wonder: mathematics educators, policy makers, and researchers. Not only are mathematics educators and researchers convinced about the importance of school mathematics, they also know how to make the case for it. If asked, they all come up with a number of reasons, and their arguments look more or less the same, whatever the cultural background of their presenters. Yet these common arguments are almost as old as school mathematics itself, and those who use them do not seem to have considered the possibility that, as times change, these arguments might have become unconvincing.

Psychologically, this attitude is fully understandable. After all, at stake is the twig on which the mathematics education community has woven its nest. And yet, as the wonderings about the status of school mathematics are becoming louder and louder, the need for a revision of our reasons can no longer be ignored. In what follows, I respond to this need by taking a critical look at some of the most popular arguments for the currently popular slogan, "Mathematics for all." This analysis is preceded by a proposal of how to think about mathematics so as to loosen the grip of clichés and to shed off hidden prejudice. It is followed by my own take on the question of what mathematics to teach, to whom, and how.

What Is Mathematics?

To justify the conviction that competence in mathematics is a condition for good citizenship, one must first address the question of what mathematics is and what role it has been playing in the life of Western society.[1] Here is a proposal: I believe that it might be useful to think about any type of human knowing, mathematics included, as an activity of, or a potential for, telling certain kinds of stories about the world. These stories may sometimes appear far removed from anything we can see or touch, but they nevertheless are believed to remain in close relationship to the tangible reality and, in the final account, are expected to mediate all our actions and improve the ways in which we are going about our human affairs. Since mathematical stories are about objects that cannot be seen, smelled, or touched, it may be a bit difficult to see that the claim of practical usefulness applies to mathematics as much as to physics or biology. But then it suffices to recall the role of, say, measurements and calculations in almost any task a person or a society may wish to undertake to realize that mathematical stories are, indeed, a centerpiece of our universal world-managing toolkit. And I have used just the simplest, most obvious example.

So, as the activity of storytelling, mathematics is not much different from any other subject taught in school. Still, its narratives are quite unlike those told in history, physics, or geography. The nature of the objects these stories are about is but one aspect of the apparent dissimilarity. The way the narratives are constructed and deemed as endorsable ("valid" or "true") makes a less obvious, but certainly not any less important, difference. It is thus justified to say that *mathematics is a discourse*—a special way of communicating, made unique by its vocabulary, visual means, routine ways of doing things, and the resulting set of endorsed narratives—of stories believed to faithfully reflect the real state of affairs. By presenting mathematics in this way (see also Sfard 2008), I am moving away from the traditional vision of mathematics as given to people by the world itself. Although definitely constrained by external reality, mathematics is to a great extent a matter of human decisions and choices, and of contingency rather than of necessity. This notion means that mathematical communication can and should be constantly monitored for its effects. In particular, nothing that regards the uses of mathematics is written in stone, and there is no other authority than us to say what needs to be preserved and what must be changed. This conceptualization, therefore, asks for a critical analysis of our common mathematics-related educational practices.

Why Mathematics? Deconstructing Some Common Answers

Three arguments for the status of mathematics as a *sine qua non* of school curricula can usually be heard these days in response to the question of why mathematics: the utilitarian, the political, and the cultural. I will call these three motives "official," to distinguish them from yet another one, which, although not any less powerful than the rest, is never explicitly stated by the proponents of the slogan "mathematics for all."

The Utilitarian Argument: Mathematics Helps in Dealing with Real-Life Problems

Let me say it again: Mathematics, just as any other domain of human knowledge, is the activity of describing—thus understanding—the world in ways that can mediate and improve our actions. It is often useful to tell ourselves some mathematical stories before we act and to repeat them as we act, while also forging some new ones. With their exceptionally high level of abstraction and the unparalleled capacity for generalization, mathematical narratives are believed to be a universal tool, applicable in all domains of our lives. And indeed, mathematics has a long and glorious history of contributions to the well-being of humankind. Ever since its inception, it has been providing us with stories that, in spite of their being concerned with the universe of intangible objects, make us able to deal with the reality around us in particularly effective and useful ways. No wonder, then, that mathematics is considered indispensable for our existence. And yet, whereas this utilitarian argument holds when the term "our existence" is understood as referring to the life of human society as a whole, it falls apart when it comes to individual lives.

I can point to at least two reasons because of which the utility claim does not work at the individual level. First, it is enough to take a critical look at our own lives to realize that we do not, in fact, need much mathematics in our everyday lives. A university professor recently said in a TV interview that in spite of his sound scientific-mathematical background he could not remember the last time he had used trigonometry, derivatives, or mathematical induction for any purpose. His need for mathematical techniques never goes beyond simple calculation, he said. As it turns out, even those whose profession requires more advanced mathematical competency are likely to say that whatever mathematical tools they are using, the tools have been learned at the job rather than in school.

The second issue I want to point to may be at least a partial explanation for the first: People do not necessarily recognize the applicability of even those mathematical concepts and techniques with which they are fairly familiar. Indeed, research of the past few decades (Brown et al. 1989; Lave 1988; and Lave and Wenger 1991) brought ample evidence that having mathematical tools does not mean knowing when and how to use them. If we ever have recourse to mathematical discourse, it is usually in contexts that closely resemble those in which we encountered this discourse for the first time. The majority of the school-learned mathematics remains in school for the rest of our lives. These days, this phenomenon is known as *situatedness of learning*, that is the dependence of the things we know on the context in which they have been learned. To sum up, not only is our everyday need for school mathematics rather limited, but also the mathematics that we could use does not make it easily into our lives. All this pulls the rug from under the feet of those who defend the idea of teaching mathematics to all because of its utility.

The Political Argument: Mathematics Empowers

Because of the universality of mathematics and its special usefulness,[2] the slogan "knowledge is power," which can now be translated into "discourses are power," applies to this special form of talk with a particular force. Ever since the advent of modernity, with its high respect for, and utmost confidence in, human reason, mathematics has been one of the hegemonic discourses of Western society. In this positivistically minded world, whatever is stated in mathematical terms tends to override any other type of argument (just recall, for instance, what counts as decisive "scientific evidence" in the eyes of the politician), and the ability to talk mathematics is thus considered an important social asset, indeed, a key to success. But the effectiveness of mathematics as a problem-solving tool is only a partial answer to the question of where this omnipotence of mathematical talk comes from. Another relevant feature of mathematics is its ability to impose linear order on anything quantifiable. Number-imbued discourses are perfect settings for decision-making and, as such, they are favored by many, and especially by politicians (and it really does not matter that all too often, politicians can only speak pidgin mathematics; the lack of competency is not an obstacle for those who know their audience and are well aware of the fact that numbers do not have to be used correctly to impress).

The second pro-math argument, one that I called political, can now be stated in just two words: Mathematics empowers. Indeed, if mathematics is the discourse of power, mathematical competency is our armor and mathematical techniques are our social survival skills. When we wonder whether mathematics is worth our effort, at stake is our agency as individuals and our independence as members of society: If we do not want to be pushed around by professional number-jugglers, we must be able to juggle numbers with them and do it equally well, if not better. Add to this idea the fact that in our society mathematics is a gatekeeper to many coveted jobs and is thus a key to social mobility, and you cannot doubt the universal need for mathematics any longer.

Now it is time for my counterarguments. The claim that "mathematics empowers" is grounded in the assumption that mathematics is a privileged discourse, a discourse likely to supersede any other. But should the hegemony of mathematics go unquestioned? On a closer look, not each of its uses may be for the good of those whose well-being and empowerment we have in mind when we require "mathematics for all." For example, when mathematics, so effective in creating useful stories about the physical reality around us, is also applied in crafting stories about children (as in "This is a below average student") and plays a decisive role in determining the paths their lives are going to take, the results may be less than helpful. More often than not, the numerical tags with which these stories label their young protagonists, rather than empowering the student, may be raising barriers that some of the children will never be able to cross. The same happens when the ability to participate in mathematical discourse is seen as a norm and the lack thereof as pathology and a symptom of a general insufficiency of the child's "potential." I will return to all this when presenting the "unofficial" argument for obligatory school mathematics. For now, the bottom line of what was written so far is simple: We need to remember that by embracing the slogan "mathematics empowers" as is, without any amendments, we may be unwittingly reinforcing social orders we wish to change. As I will be arguing in the concluding part of this editorial, trying to change the game may be much more "empowering" than trying to make everybody join in and play it well.

The Cultural Argument: Mathematics Is a Necessary Ingredient of Your Cultural Makeup

In the latest paragraph, I touched upon the issue of the place of mathematics in our culture and in an individual person's identity. I will now elaborate on this topic while presenting the cultural argument for teaching mathematics to all.

Considering the fact that to think means to participate in some kind of discourse, it is fair to say that our discourses, those discourses in which each of us is able to participate, constitute who we are as social beings. In the society that appreciates intellectual skills and communication, the greater and more diverse our discursive repertoire, the richer, more valued, and more attractive our identities. However, not all discourses are made equal, so the adjective "valued" must be qualified. Some forms of communicating are considered to be good for our identities, and some others much less so. As to mathematics, many would say that it belongs to the former category. Considered as a pinnacle of human intellectual achievement and thus as one of the most precious cultural assets, it bestows some of its glory even on peripheral members of the mathematical community. Those who share this view believe that mathematical competency makes you a better person, if only because of the prestigious membership that it affords. A good illustration of this claim comes from an Israeli study (Sfard and Prusak 2005) in which 16-year-old immigrant students, originally from the former Soviet Union, unanimously justified their choice of the advanced mathematics program with claims that mathematics is an indispensable ingredient of one's identity. "Without mathematics, one is not a complete human being," they claimed.

But the truth is that the attitude demonstrated by those immigrant students stands today as an exception rather than a rule. In the eyes of today's young people, at least those who come from cultural backgrounds with which I am well acquainted, mathematics does not seem to have the allure it had for my generation. Whereas this statement can be supported with numbers that show a continuous decline in percentages of graduates who choose to study mathematics (or science)—and currently, this seems to be a general trend in the Western world[3]—I can also present some firsthand evidence. In the same research in which the immigrant students declared their need for mathematical competency as a necessary ingredient of their identities, the Israeli-born participants spoke about mathematics solely as a stepping stone for whatever else they would like to do in the future. Such an approach means that one can dispose with mathematics once it has fulfilled its role as an entrance ticket to preferable places. For the Israeli-born participants, as for many other young people these days, mathematical competency is no longer a highly desired ingredient of one's identity.

Considering the way the world has been changing in the past few decades, it may not be too difficult to account for this drop in the popularity of mathematics. One of the reasons may be the fact that mathematical activity does not match the life experiences typical of our postmodern communication-driven world. As aptly observed in a recent book by Susan Cain (2012), the hero of our times is a vocal, assertive extrovert with well-developed communicational skills and an insatiable appetite for interpersonal contact. Although there is a clear tendency, these days, to teach mathematics in collaborative groups—the type of learning that is very much in tune with this general trend toward the collective and the interpersonal—we need to remember that one cannot turn mathematics into a discourse-for-oneself unless one also *practices* talking mathematics to oneself. And yet, as long as interpersonal communication is the name of the game and a person with a preference for the intrapersonal dialogue risks marginalization, few students may be ready to suspend their intense exchanges with others for the sake of well-focused, time consuming conversation with themselves.

In spite of all that has been said above, I must confess that the cultural argument is particularly difficult for me to renounce. I have been brought up to love mathematics for what it is. Born into the modernist world ruled by logical positivism, I believed that mathematics must be treated as a queen even when it acts as a servant. Like the immigrant participants of Anna Prusak's study, I have always felt that mathematics is a valuable, indeed indispensable, ingredient of my identity—an element to cherish and of which to be proud. But this belief is just a matter of emotions. Rationally, there is little I can say in defense of this stance. I am acutely aware of the fact that times change and that, these days, modernist romanticism is at odds with postmodernist pragmatism. In the end, I must concede that the designation of mathematics as a cultural asset is not any different than that of poetry or art. Thus, however we look at it, the cultural argument alone does not justify the prominent presence of mathematics in school curricula.

The Unofficial Argument: Mathematics Is a Perfect Selection Tool

My last argument harks back to the abuses of mathematics to which I hinted while reflecting on the statement "mathematics empowers." I call it unofficial because no educational policy maker would admit to its being the principal, if not the only, motive for his or her decisions. I am talking here about the use of school mathematics as a basis for the measuring-and-labeling practices mentioned above. In our society, grades in mathematics serve as one of the main criteria for selecting school graduates for their future careers. Justifiably or not, mathematics is considered to be the *lingua franca* of our times, the universal language, less sensitive to culture than any other well-defined discourse. No intellectual competency, therefore, seems as well suited as mathematics for the role of a universal yardstick for evaluating and comparing people. Add to this the common conviction that "Good in math = generally brilliant" (with the negation being, illogically, "not good in math = generally suspect"), and you begin realizing that teaching mathematics and then assessing the results may be, above all, an activity of classifying people with "price tags" that, once attached, will have to be displayed whenever a person is trying to get access to one career or another. I do not think that an elaborate argument is needed to deconstruct this kind of motive. The very assertion that this harmful practice is perhaps the only reason for requiring mathematics for all should be enough to make us rethink our policies.

What Mathematics and Why? A Personal View

It is time for me to make a personal statement. Just in case I have been misunderstood, let me make it clear: I do care for mathematics, and I am as concerned as anybody about its future and the future of those who are going to need it. All that I said above grew from this genuine concern. By no means do I advocate discontinuing the practice of teaching mathematics in school. All I am trying to say is that we should approach the task in a more flexible, less authoritarian way, while giving more thought to the question of how much should be required from all and how much choice should be left to the learner. In other words, I propose that we rethink school mathematics and revise it quite radically. As I said before, if there is a doubt about the game being played, let us change this game rather than just trying to play it well. These days, deep, far-reaching change is needed in what we teach, to whom, and how.

I do have a concrete proposal with regard to what we can do, but let me precede this discussion with two basic "don't"s. First, let us not use mathematics as a universal instrument for selection. This practice hurts the student, and it spoils the mathematics that is being learned. Second, let us not force the traditional school curriculum on everybody, and, whatever mathematics we do decide to teach, let us teach it in a different way.

In the rest of this editorial, let me elaborate on this latter issue, which, in more constructive terms, can be stated as follows: Yes, let us teach everybody *some* mathematics, the mathematics whose everyday usefulness is beyond question. Arithmetic? Yes. Some geometry?

Definitely. Basic algebra? No doubt. Add to this some rudimentary statistics, the extremely useful topic that is still only rarely taught in schools, and the list of what I consider as "mathematics for all" is complete. And what about trigonometry, calculus, liner algebra? Let us leave these more advanced topic as electives, to be chosen by those who want to study them.

But the proposed syllabus does not, per se, convey the idea of the change I had in mind when claiming the need to rethink school mathematics. The question is not just of what to teach or to whom, but also of how to conceptualize what is being taught so as to make it more convincing and easier to learn. There are two tightly interrelated ways in which mathematics could be framed in school as an object of learning: We can think about mathematics as the *art of communicating* or as one of the basic forms of *literacy*. Clearly, both these framings are predicated on the vision of mathematics as a discourse. Moreover, a combination of the two approaches could be found so that the student can benefit from both. Let me briefly elaborate on each one of the two framings.

Mathematics as the Art of Communicating

As a discourse, mathematics offers special ways of communicating with others and with oneself. When it comes to the effectiveness of communication, mathematics is unrivaled: When at its best, it is ambiguity-proof and has an unparalleled capacity for generalization. To put it differently, mathematical discourse appears to be infallible—any two people who follow its rules must eventually agree, that is, endorse the same narratives; in addition, this discourse has an exceptional power of expression, allowing us to say more with less.

I can see a number of reasons why teaching mathematics *as the art of communicating* may be a good thing to do. First, it will bring to the fore the interpersonal dimension of mathematics: The word *communication* reminds us that mathematics originates in a conversation between mathematically minded thinkers, concerned about the quality of their exchange at least as much as about what this exchange is all about. Second, the importance of the communicational habits one develops when motivated by the wish to prevent ambiguity and ensure consensus exceeds the boundaries of mathematics. I am prepared to go so far as to claim that if some of the habits of mathematical communication were regulating all human conversations, from those that take place between married couples to those between politicians, our world would be a happier place to live. Third, presenting mathematics as the art of interpersonal communication is, potentially, a more effective educational strategy than focusing exclusively on intrapersonal communication. The interpersonal approach fits with today's young people's preferences. It is also easier to implement. After all, shaping the ways that students talk to each other is, for obvious reasons, a more straightforward job than trying to mold their thinking directly. Fourth, framing the task of learning mathematics as perfecting one's ability to communicate with others may be helpful, even if not sufficient, in overcoming the situatedness of mathematical learning. Challenging students to find solutions that would convince the worst skeptic will likely help them develop the lifelong habit of paying attention to the way they talk (and thus think!). This kind of attention, being focused on one's own actions, may bring about discursive habits that are less context-dependent and more universal than those that develop when the learner is almost exclusively preoccupied with mathematical objects. There may be more, but I think these four reasons should suffice to explain why teaching mathematics as an art of communication appears to be a worthy endeavor.

Mathematics as a Basic Literacy

While teaching mathematics as an art of communicating, we stress the question of *how* to talk. Fostering mathematical literacy completes the picture by emphasizing the issues of *when* to talk mathematically and *what about*.

Although, nowadays, mathematical literacy is a buzz phrase, a cursory review of literature suffices to show that there is not much agreement on how it should be used. For the sake of the present conversation, I define mathematical literacy as the ability to decide not just about *how* to participate in mathematical discourse but also about *when* to do so. The emphasis on the word *when* signals that mathematical literacy is different from the type of formal mathematical knowledge that is being developed, in practice if not in principle, through the majority of present-day curricula. These curricula offer mathematics as, first and foremost, a self-sustained discourse that speaks about its own unique objects and has few ties to anything external. Thus, they stress the *how* of mathematics to the neglect of the *when*. Mathematical literacy, in contrast, means the ability to engage in mathematical communication whenever this approach may help in understanding and manipulating the world around us. It thus requires fostering the *how* and the *when* of mathematical routines at the same time. To put it in discursive terms, along with developing students' participation in mathematical discourse, we need to teach them how to combine this discourse with other ones. Literacy instruction must stress students' ability to switch to mathematical discourse from any other discourse whenever appropriate and useful, and it has to foster the capacity for incorporating some of the metamathematical rules of communication into other discourses.

My proposal, therefore, is to replace the slogan "mathematics for all" with the call for "mathematical literacy for all." Arithmetic, geometry, elementary algebra, the basics of statistics—these are mathematical discourses that, I believe, should become part and parcel of every child's literacy kit. This ideal is easier said than done, of course. Because of the inherent situatedness of learning, the call for mathematical literacy presents educators with a major challenge. The question of how to teach for mathematical literacy must be theoretically and empirically studied. When we consider the urgency of the issue, we should make sure that such research is given high priority.

In this editorial, I tried to make the case for a change in the way we think about school mathematics. In spite of the constant talk about reform, the current mathematical curricula are almost the same in their *content* (as opposed to pedagogy) as they were decades, if not centuries, ago. Times change, but our general conception of school mathematics remains invariant. As mathematics educators, we have a strong urge to preserve the kind of mathematics that has been at the center of our lives ever since our own days as school students. We want to make sure that the new generation can have and enjoy all those things that our own generation has seen as precious and enjoyable. But times do change, and students' needs and preferences change with them. With the advent of knowledge technologies that allow an individual to be an agent of her or his own learning, our ability to tell the learner what to study changes as well. In this editorial, I proposed that we take a good look at our reasons and then, rather than imposing one rigid model on all, restrict our requirements to a basis from which many valuable variants of mathematical competency may spring in the future.

Notes

1. I am talking about Western society because due to my personal background, this is the only one I feel competent to talk about. The odds are, however, that in our globalized world there is not much difference, in this respect, between Western society and all the others.

2. Just to make it clear, the former argument that mathematics is not necessarily useful in every person's life does not contradict the claim about its general usefulness!

3. As evidenced by numerous publications on the drop in enrollment to mathematics-related university subjects (e.g., Garfunkel and Young 1998; ESRC 2006; and OECD 2006) and by the frequent calls for research projects that examine ways to reverse this trend (see, e.g., the Targeted Initiative on Science and Mathematics Education in the United Kingdom, http://tisme-scienceandmaths.org/), the decline in young people's interest in mathematics and science is generally considered these days as one of the most serious educational problems, to be studied by educational researchers and dealt with by educators and policy makers.

References

Brown, J. S., Collins, A., and Duguid, P. (1989). Situated cognition and the culture of learning. *Educational Researcher, 18*(1), 32–42.

Cain, S. (2012). *Quiet—The power of introverts in a world that can't stop talking.* New York: Crown Publishers.

Economic and Social Research Council (ESRC), Teaching and Learning Research Programme. (2006). *Science education in schools: Issues, evidence and proposals.* Retrieved from http://www.tlrp.org/pub/documents/TLRP_Science_Commentary_FINAL.pdf.

Garfunkel, S. A., and Young, G. S. (1998). The Sky Is Falling. *Notices of the AMS, 45*, 256–257.

Lave, J. (1988). *Cognition in practice.* New York: Cambridge University Press.

Lave, J., and Wenger, E. (1991). *Situated learning: Legitimate peripheral participation.* New York: Cambridge University Press.

Organization for Economic Co-operation and Development, Global Sciences Forum. (2006). *Evolution of student interest in science and technology studies.* Retrieved from http://www.oecd.org/dataoecd/16/30/36645825.pdf.

Sfard, A. (2008). *Thinking as communicating: Human development, the growth of discourses, and mathematizing.* Cambridge, U.K.: Cambridge University Press.

Sfard, A., and Prusak, A. (2005). Telling identities: In search of an analytic tool for investigating learning as a culturally shaped activity. *Educational Researcher, 34*(4), 14–22.

Math Anxiety: Who Has It, Why It Develops, and How to Guard against It

Erin A. Maloney and Sian L. Beilock

Understanding Math Anxiety

For people with math anxiety, opening a math textbook or even entering a math classroom can trigger a negative emotional response, but it does not stop there. Activities such as reading a cash register receipt can send those with math anxiety into a panic. Math anxiety is an adverse emotional reaction to math or the prospect of doing math [1]. Despite normal performance in most thinking and reasoning tasks, people with math anxiety perform poorly when numerical information is involved.

Why is math anxiety tied to poor math performance? One idea is that math anxiety is simply a proxy for low math ability, meaning that individuals with math anxiety are less skilled or practiced at math than their nonanxious counterparts. After all, math anxious individuals tend to stay away from math classes and learn less math in the courses they do take [1]. Yet low math ability is not the entire explanation for why math anxiety and poor math performance co-occur. It has been shown that people's anxiety about doing math—over and above their actual math ability—is an impediment to math achievement [2]. When faced with a math task, math anxious individuals tend to worry about the situation and its consequences. These worries compromise cognitive resources, such as working memory, a short-term system involved in the regulation and control of information relevant to the task at hand [3]. When the ability of working memory to maintain task focus is disrupted, math performance often suffers.

Despite the progress made in understanding how math anxiety relates to math performance, only limited attention has been devoted to the antecedents of math anxiety. Determining who is most likely to develop math anxiety, when they develop it, and why is essential for gaining a full understanding of the math anxiety phenomenon and its role in math achievement.

Math Anxiety: Antecedents and Developmental Trajectory

The first years of elementary school are critical for learning basic mathematical skills. Yet until recently the dominant view among educators and researchers alike was that math anxiety only arose in the context of complex mathematics (e.g., algebra) and thus was not present in young children. Math anxiety was thought to develop in junior high school, coinciding with the increasing difficulty of the math curriculum toward the end of elementary school [2]. Recent research challenges this assumption. Not only do children as young as first grade report varying levels of anxiety about math, which is inversely related to their math achievement [4], but also this anxiety is also associated with a distinct pattern of neural activity in brain regions associated with negative emotions and numerical computations. When performing mathematical calculations, math anxious children, relative to their less anxious counterparts, show hyperactivity in the right amygdala regions that are important for processing negative emotions. This increased amygdala activity is accompanied by reduced activity in brain regions known to support working memory and numerical processing (e.g., the dorsolateral prefrontal cortex and posterior parietal lobe) [5].

Both social influences and cognitive predispositions probably play a role in the onset of math anxiety in early elementary school. In terms of social influences, teachers who are anxious about their own math abilities impart these negative attitudes to some of their students. Interestingly, this transmission of negative math attitudes seems to fall along gender lines. Beilock and colleagues found that it was only the female students of highly math anxious female teachers (>90% of elementary teachers in the United States are female) who tended to endorse the stereotype that "boys are good at math, girls at reading" by the end of a school year. Girls who endorsed this stereotype were also most likely to be behind in math at the end of the school year [6]. Similar to how social mores are passed down from one generation to another, negative math attitudes seem to be transmitted from teacher to student.

Some children may also have a cognitive predisposition to develop math anxiety. In adults, math anxiety is associated with deficits in one or more of the fundamental building blocks of mathematics. For example, adults who are math anxious are worse than their nonanxious peers at counting objects, at deciding which of two numbers represents a larger quantity, and at mentally rotating 3-D objects [7–9]. Similar to how people who lack knowledge in a particular domain are often easily swayed by negative messages [10], children who start formal schooling with deficiencies in these mathematical building blocks may be especially predisposed to pick up on social cues (e.g., their teacher's behavior) that highlight math in negative terms.

Alleviating Math Anxiety

Understanding the antecedents of math anxiety provides clues about how to prevent its occurrence. For instance, bolstering basic numerical and spatial processing skills may help to reduce the likelihood of developing math anxiety. If deficiencies in basic mathematical competencies predispose students to becoming math anxious, then early identification of at-risk students (coupled with targeted exercises designed to boost their basic mathematical competencies and regulate their potential anxieties) may help to prevent children from developing math anxiety in the first place.

Knowledge about the onset of math anxiety also sheds light on how to weaken the link between math anxiety and poor math performance in those who are already math anxious. If exposure to negative math attitudes increases the likelihood of developing math anxiety, which in turn adversely affects math learning and performance, then regulation of the negativity associated with math situations may increase math success, even for those individuals who are chronically math anxious. Support for this idea comes from work showing that when simply anticipating an upcoming math task, math anxious individuals who show activation in a frontoparietal network known to be involved in the control of negative emotions perform almost as well as their nonanxious counterparts on a difficult math test [11]. These neural findings suggest that strategies that emphasize the regulation and control of negative emotions—even before a math task begins—may enhance the math performance of highly math anxious individuals.

One means by which people can regulate their negative emotions is expressive writing in which people are asked to write freely about their emotions for 10–15 minutes with respect to a specific situation (e.g., an upcoming math exam). Writing is thought to alleviate the burden that negative thoughts place on working memory by affording people an opportunity to re-evaluate the stressful experience in a manner that reduces the necessity to worry altogether. Demonstrating the benefits of expressive writing, Ramirez and Beilock showed that having highly test anxious high school students write about their worries before an upcoming final exam boosted their scores from B– to B+ (even after taking into account grades across the school year) [12]. Similar effects have been found specifically for math anxiety. Writing about math-related worries boosts the math test scores of math anxious students [13].

Negative thoughts and worries can also be curtailed by reappraisal, or reframing, techniques. Simply telling students that physiological responses often associated with anxious reactions (e.g., sweaty palms, rapid heartbeat) are beneficial for thinking and reasoning can improve test performance in stressful situations [14]. Having students think positively about a testing situation can also help them to reinterpret their arousal as advantageous rather than debilitating. For example, when students view a math test as a challenge rather than a threat, the stronger their physiological response to the testing situation (measured here in terms of salivary cortisol), the better, not worse, is their performance [15].

Summing Up

Education, psychology, and neuroscience researchers have begun to uncover the antecedents of math anxiety. Not only is math anxiety present at the beginning of formal schooling, which is much younger than was previously assumed, but its development is also probably tied to both social factors (e.g., a teacher's anxiety about her own math ability) and a student's own basic numerical and spatial competencies—where deficiencies may predispose students to pick up on negative environmental cues about math. Perhaps most striking, many of the techniques used to reduce or eliminate the link between math anxiety and poor math performance involve addressing the anxiety rather than training in math itself. When anxiety is regulated or reframed, students often see a marked increase in their math performance. These findings underscore the important role that affective factors play in situations that require mathematical reasoning. Unfortunately, it is still quite rare that numerical cognition research takes into account issues of math anxiety when studying numerical and mathematical processing. By ignoring the powerful role that anxiety plays in mathematical situations, we are overlooking an important piece of the equation in terms of understanding how people learn and perform mathematics.

Acknowledgments

This work was supported by U.S. Department of Education, Institute of Education Sciences Grant R305A110682 and NSF CAREER Award DRL-0746970 to Sian Beilock as well as the NSF Spatial Intelligence and Learning Center (grant numbers SBE-0541957 and SBE-1041707).

References

[1] Hembree, R. (1990) The nature, effects, and relief of mathematics anxiety. *J. Res. Math. Educ.* 21, 33–46.

[2] Ashcraft, M. H., et al. (2007) Is math anxiety a mathematical learning disability? In *Why Is Math So Hard for Some Children? The Nature and Origins of Mathematical Learning Difficulties and Disabilities* (Berch, D. B., and Mazzocco, M. M. M., eds.), pp. 329–348, Brookes.

[3] Beilock, S. L. (2010) *Choke: What the Secrets of the Brain Reveal about Getting It Right When You Have To*, Simon & Schuster.

[4] Ramirez, G., et al. (2012) Math anxiety, working memory and math achievement in early elementary school. *J. Cogn. Dev.* 14.2, 187–202.

[5] Young, C. B., et al. (2012) Neurodevelopmental basis of math anxiety. *Psychol. Sci.* 23, 492–501.

[6] Beilock, S. L., et al. (2010) Female teachers' math anxiety affects girls' math achievement. *Proc. Natl. Acad. Sci. U.S.A.* 107, 1060–1063.

[7] Maloney, E. A., et al. (2010) Mathematics anxiety affects counting but not subitizing during visual enumeration. *Cognition* 114, 293–297.

[8] Maloney, E. A., et al. (2011) The effect of mathematics anxiety on the processing of numerical magnitude. *Q. J. Exp. Psychol.* 64, 10–16.

[9] Maloney, E. A., et al. (2012) Reducing the sex difference in math anxiety: The role of spatial processing ability. *Learn. Individ. Diff.* 22, 380–384.

[10] Petty, R. E., and Cacioppo, J. T. (1986) The elaboration likelihood model of persuasion. *Adv. Exp. Soc. Psychol.* 19, 123–205.

[11] Lyons, I. M., and Beilock, S. L. (2011) Mathematics anxiety: Separating the math from the anxiety. *Cereb. Cortex* http://dx.doi.org/10.1093/cercor/bhr289.

[12] Ramirez, G., and Beilock, S. L. (2011) Writing about testing worries boosts exam performance in the classroom. *Science* 331, 211–213.

[13] Park, D., et al. (2011) Put your math burden down: expressive writing for the highly math anxious. Paper presentation at the Midwestern Psychology Association, Chicago.
[14] Jamieson, J. P., et al. (2010) Turning the knots in your stomach into bows: Reappraising arousal improves performance on the GRE. *J. Exp. Soc. Psychol.* 46, 208–212,
[15] Mattarella-Micke, A., et al. (2011) Choke or thrive? The relation between salivary cortisol and math performance depends on individual differences in working memory and math anxiety. *Emotion* 11, 1000–1005.

How Old Are the Platonic Solids?

DAVID R. LLOYD

Recently a belief has spread that the set of five Platonic solids has been known since prehistoric times, in the form of carved stone balls from Scotland, dating from the Neolithic period. A photograph of a group of these objects has even been claimed to show mathematical understanding of the regular solids a millennium or so before Plato. I argue that this is not so. The archaeological and statistical evidence do not support this idea, and it has been shown that there are problems with the photograph. The high symmetry of many of these objects can readily be explained without supposing any particular mathematical understanding on the part of the creators, and there seems to be no reason to doubt that the discovery of the set of five regular solids is contemporary with Plato.

Introduction

The flippant and rather too obvious answer to the question in the title is "as old as Plato." However, a little investigation shows that the question is more complex and needs some firming up. If it is taken as referring to *any* of the solids, then there is ancient testimony (Heath 1981, 162) that some of the five predate Plato. This tradition associates the tetrahedron, cube, and dodecahedron with the Pythagoreans.[1] The same source (Heath 1981, 162) claims that the remaining two, the octahedron and icosahedron, were actually discoveries made by Plato's collaborator Theaetetus, and although there have been doubts about this claim, particularly for the octahedron (Heath 1956, 438), there is now substantial support for the idea that this ancient testimony, giving major credit to Theaetetus, is reliable (Waterhouse 1972; Artmann 1999, 249–251, 285, 296–298, 305–308).

Nevertheless, Plato is responsible for giving us the first description we have of the complete set of the five regular solids, in his dialogue *Timaeus* (about 360 BCE), and probably this is the reason we speak of Platonic solids. Plato was not claiming any originality here; rather he seemed to be assuming that the hearers in his dialogue, and his readers, would already be familiar with these solids. For Plato, their importance was that these were the "most beautiful" (*kallistos*) structures possible, and that from these he could construct a theory of the nature of matter. It is probable that this *kallistos* description alludes to what Waterhouse (1972) claims was then a relatively new discovery, that of the concept of regularity, and to the demonstration, which is recorded later by Euclid and which is probably also due to Theaetetus (Waterhouse 1972; Artmann 1999, 249–251), that there can only be five such structures.[2] Thus, at least until recently, the consensus has been that although Plato is not responsible for the solids themselves, the discovery of the *set* of five, and of the fact that there are only five, are contemporary with him.

However, a belief has spread recently that the complete set of these five regular solids has been known since much earlier times, in the form of decorated stone balls from Scotland, all from the Neolithic period. The origin of this belief is a photograph of five of these objects (Figure 1), which first appeared in a book published more than 30 years ago (Critchlow 1979). In this book, it was alleged that the discovery of the set of the five regular solids, and at least some of the associated theory, predates Plato and his collaborators by a millennium[3]; other claims for ancient mathematical knowledge also appear in this book.[4]

The claim that the five solids were known at this time was accepted by Artmann, who comments that "All of the five regular solids appear in these decorations" (Artmann 1999, 300–301), though he is very skeptical of the other claims. He notes that a photograph of five of the objects had appeared in *Mathematics Teaching* (1985, p. 56). More recently, Atiyah and Sutcliffe (2003) published a paper that reproduces the photograph shown by Critchlow, with a caption describing these objects as "models" of the five solids, but they gave no attribution for the source of the illustration. This paper has also appeared in the collected works of Atiyah (2004), and in a review of this collection (Pekkonen 2006), it was claimed that the stone balls "provide perfect models of the Platonic Solids one thousand years before Plato or Pythagoras." The word "models" suggests some sort of mathematical understanding on the

FIGURE 1. Five stone balls, decorated with tapes that pick out some of the symmetry elements (copyright G. Challifour).

part of the makers of the objects, and, almost certainly unintentionally, echoes Critchlow's earlier claims. It is probably the appearance of the photograph in the paper by Atiyah and Sutcliffe, and the authority of the authors, which is responsible for the recent growth within the scientific and mathematical communities of the idea that the set of the five Platonic solids was known long before Plato. The photograph (see Figure 1) now makes occasional appearances in lectures on aspects of polyhedra and their symmetries, as an illustration of the idea, which can also be found on various websites, that the solids are much older than had been thought until recently.

Clearly the appearance of the *set* of all five solids at this early time, if true, would be quite remarkable, and should the suggestion of some sort of mathematical understanding be justifiable, there would be substantial implications for the history of mathematics. I examine some aspects of the story in more detail here.

Archaeology and Statistical Frequencies

It is important to emphasize at the outset that the objects shown in Figure 1 are genuine Neolithic artifacts, almost certainly from the large collection belonging to The National Museum of Scotland (NMS),[5] which the museum dates to within the period 3200 BCE to 2500 BCE. More than 400 such objects are known (Edmonds 1992), almost all roughly spherical, though there are a few prolate spheroids, and most are carved with approximately symmetric arrangements. With few exceptions, these objects are small enough to be held comfortably in one hand. The principal study of them, with a complete listing of those known at the time, is by Marshall (1977, 40–72). On many but not all of these objects, the carving is quite deep, so that there is a set of protrusions or knobs with a common center, as in the examples in Figure 1. They have been found in a variety of locations, but almost all within the boundaries of present-day Scotland.

Of the five shown in Figure 1, only the second and fifth types (reading from the left) have been discovered in large numbers. It is tempting to classify these two types as "tetrahedral" and "octahedral." However, such terms are not used by archaeologists, who normally call these "4-knob" and "6-knob" forms. This label is sometimes abbreviated to "4K" and "6K," and this convention is used here. The symmetries of the objects are discussed below. Some 43 of the 4K objects, and 173 of the 6K, are known. The other numbers given by Marshall, up to 16K, are 3K (6), 5K (2),[6] 7K (18), 8K (10), 9K (3), 10K (4), 11K (1), 12K (8), 13K (0), 14K (5), 15K (1), and 16K (1). Although there were carvers who were prepared to tackle spheres with larger numbers of knobs,[7] perhaps to demonstrate their skill, most seem to have preferred to deal with 14K or fewer. Given that these objects could only be created by slow hand grinding, perhaps with the help of leather straps and abrasive scraps of rock, this preference is not surprising; it must be difficult, if not impossible, to achieve any precision when the knobs become small.

The first object in Figure 1, with 8K, is an example of a rare species. Although 10 examples of 8K are known, by no means all of these have the cube form shown here. Marshall, who had personally examined almost all the known examples, notes that for the 8K objects, "This group of balls has variety in the disposition of the eight knobs" (Marshall 1977, 42). No such comment is made about any other group, and examination of published illustrations of the 8K objects shows a total of only three that have this approximately cubic shape.[8] There are substantially more (18) examples known for the 7K version, so it does not seem that the cube form of the 8K objects held a particular significance for the makers or users.

There is a small peak in the numbers at 12K and 14K, suggesting that these were interesting to the carvers. However, purely on statistical grounds, it seems unlikely that the makers of such a wide variety of objects would have made any connection to the particular selection shown in Figure 1, since two of these are commonly found, while the others are all unusual. The choice of this particular set of five for the photograph is probably dependent on knowledge that the carvers did not possess.

The Provenance of the Objects in Figure 1

The illustration is taken from Critchlow (1979, 132, figure 114), who believes that he can demonstrate a high level of mathematical skills among the Neolithic peoples of the British Isles. The carved stone balls are only part of this narrative. The original caption to the photograph reads "A full set of Scottish Neolithic 'Platonic solids'—a millennium before Plato's time."

Oddly, the book gives no indication of which museum owns the five objects shown, and because the text associated with this illustration gives a detailed description of the five balls that belong to the Ashmolean Museum at Oxford, some have concluded, incorrectly, that Figure 1 shows objects from the Ashmolean collection (Lawlor 1982; Atiyah and Sutclifffe 2003). It is evident from Critchlow's text that he is describing a completely different set, with different numbers of knobs, from that shown in Figure 1. The Ashmolean set can be seen on the web, together with the original drawings made by a curator at the time of acquisition,[9] and it is clearly different from Figure 1.

It is therefore of some interest to try to establish which objects *were* photographed, if only to see whether they might have been found together, since that could imply that the carvers or owners saw some sort of connection among these objects. Accompanying the original of Figure 1 there is another photograph (Critchlow 1979, 132, figure 113), apparently taken on the same occasion, of a group of 4K objects, several of which carry markings that identify them as belonging to what was then the Scottish National Museum of Antiquities (NMA in the Marshall list), now The National Museum of Scotland (NMS) in Edinburgh. The fourth object in Figure 1 is also shown in a close-up view (Critchlow 1979, 149, figure 146), with the caption "on show in Edinburgh." Comparison of this object with one of the illustrations on the NMS website[10] makes it very likely that these two photographs are of the same object, recorded as having been discovered in Aberdeenshire. Thus it seems almost certain that the five objects in Figure 1 are from the NMS.[11]

There are so many examples of 6K objects that it is not possible to identify which particular one appears at the extreme right of the photograph. However, inspection of the images of the 8K objects on the NMS website shows that the leftmost object in Figure 1, the cube, is almost certainly one that was found in Ardkeeling, Strypes, Moray.[12]

From comparison with the group of 4K objects shown by Critchlow (1979, 132), and the museum's own photograph, the second object in Figure 1 (4K) is probably one[13] that is described as of "unknown origin." Finally, the fourth object in Figure 1, like the third object, is recorded as having been discovered in Aberdeenshire.[14] However, unlike the third object, this one is noted as having belonged to a John Rae (who died in 1893).

The museum descriptions vary considerably for these objects and indicate that they have been accumulated from four or five different sources over the years. Thus the five could not have been found together; they were brought together in the photograph, probably for the first time, in order to support a particular view.

How Many Platonic Solids?

Figure 1 was claimed, and has been widely accepted, as evidence that the carvers were familiar with all *five* of the Platonic solids. However, there is a problem with the tapes in this figure, which are not used consistently. For the second, fourth, and fifth objects, they are used to connect the knobs, though for the second one (the tetrahedron), connections are also shown between the interstices. However, for the first object, the tapes are connecting four-fold interstitial positions, and for the third object, they connect the three-fold interstices. This inconsistency has tended to hide the fact, first pointed out by Hart (1998) and later by le Bruyn (2009),[15] that Figure 1 actually includes only *four* of the Platonic structures. In his text, Critchlow defines the knobs as vertices of the Platonic solids; this convention agrees with that used by chemists in describing atom positions in molecules. According to this convention, Figure 1 shows, from left to right, a cube, a tetrahedron, two icosahedra, and an octahedron. Although on the third and fourth objects the tapes are arranged differently, each is a 12K object. The main difference between these two is in the form of the knobs, one pronounced and well-rounded, the other flatter, so that the eye picks out pentagons, at least when helped by the tapes added to this object. If the photographs on the museum's website, without tapes, are compared, then the difference between the two is much less obvious.

Thus there is no evidence here for a structure related to the fifth Platonic solid. If such a structure were to exist, it would be a 20K object with the knobs at the corners of a dodecahedron. There is a reference to "the dodecahedron on one specimen in the Museum of Edinburgh" (Artmann 1999, 305), but since this occurs in a paragraph which mentions the photographs in Critchlow, this almost certainly means the close-up photograph of the third object in Figure 1, with only 12K, which was referred to above (Critchlow 1979, 149, figure 146).

According to the Marshall list, there are only two balls known that have 20K; one of these balls is at the NMS. Alan Saville, senior curator for earliest prehistory at this museum, has provided a photograph which shows that this object is complex, and certainly not a dodecahedron. It could be considered as a modified octahedron, with five large knobs in the usual positions, but with the sixth octahedral position occupied by 12 small knobs, and in addition there are also three small triangles carved at some of the interstices, the three-fold positions of the "octahedron." These bumps make up a total of 20 "protrusions," though the word "knobs" is hard to justify.

The other 20K object is at the Kelvingrove Art Gallery and Museum in Glasgow. Photographs taken by Tracey Hawkins, assistant curator, show that this object also is far from being a dodecahedron, though this time there are 20 clearly defined knobs of roughly the same size. The shape is somewhat irregular, but two six-sided pyramids can be picked out, and much of the structure, though not all, is deltahedral in form, with sets of three balls at the corners of equilateral triangles.

No dodecahedral form with 20K has yet been found. There is therefore no evidence that the carvers were familiar with all five Platonic solids. Even for the four that they did create, there is nothing to suggest that they would have thought these in any way different from the multitude of other shapes that they carved. Thus there is no evidence that the concept of the set of five regular solids predates Plato. Nevertheless, the carvers have come up with a variety of interesting shapes, almost all with high symmetries, and I now turn to an examination of a wider grouping of these objects and suggest how the symmetries may have arisen.

The High Symmetries of These Ancient Objects

The five objects in Figure 1 show, at least approximately, the symmetries of four of the Platonic solids. However, most of the other known carved balls up to 14K, and many of the larger ones, also have approximations to high symmetries. Among the balls with relatively small numbers of knobs, up to 14K, almost all have deltahedral structures, formed by the linking of equilateral triangles. (A few of the balls have become worn or damaged, so the original form is not always obvious.)

Even though there is no evidence for the full set of Platonic solids among these objects, it might still be claimed that their symmetries suggest some sort of mathematical competence at this time. Thus it seems worthwhile to enquire if there could be other, nonmathematical reasons why such symmetric structures might have been created, but any such attempt at "explanation" needs to be as simple as possible.

The carvers are likely to have wanted to make their carved knobs as distinct as possible, and since close grinding work is required, using only the simplest of technologies, there would have been a need to keep the knobs as far away from each other as possible, simply to have room to work. Also, they would probably have seen the equilateral triangles that are generated automatically by packing in a pile of roughly spherical pebbles, or of fruits. They would have been familiar with seed heads of plants, where similar packing effects can be seen over part of a sphere, and it does not need much imagination to try to extend these triangular patterns to cover a complete sphere, which would generate a deltahedral structure.

If we assume that a guiding principle for the carvers was to keep the carved knobs as far away from each other as possible, or, equivalently, to produce an even spacing between these knobs, then the structures can be modeled in terms of repulsion between points on a sphere. Within chemistry, there is a well-known qualitative approach called "electron pair repulsion theory." This theory rationalizes the observed geometry of simple molecules as the result of minimizing interactions between bond (and other) electron pairs by maximizing the distances between them, and in the commonest cases of four and six pairs, tetrahedral and octahedral molecular geometries are often rationalized in these terms. However, the numbers of knobs in these objects span a far greater range than that found for electron pairs in molecules. A much more useful set of data for the present comparison is available from calculations on the rather similar classical problem of the distribution of *N* electric charges over the surface of a sphere (Erber and Hockney 1997; Atiyah and Sutcliffe 2003). This problem is sometimes referred to as the Thomson problem (Atiyah and Sutcliffe 2003); an alternative name is the surface Coulomb problem (Erber and Hockney 1997).

Encouragingly for chemists, the results of such calculations of the minimum energy configuration of the *N* charges accurately reproduce the predictions of electron pair repulsion theory, as *N* varies over the normal range of numbers of electron pairs found in molecules. They show an almost total preference for deltahedra as the minimum energy configurations, and it is noticeable that the majority of the carved stone balls are also deltahedra, though for balls with large numbers of knobs, other patterns appear.

However, the only Platonic solids that are generated by minimizing interactions over the surface of a sphere are the three that are also deltahedra. The cube is not a minimum energy configuration; $N = 8$ gives the square antiprism, in which opposite faces of a cube have been rotated against each other by 45°. Also, the dodecahedron is not a minimum; the calculation for 20 particles shows a deltahedron with threefold symmetry.

The calculations become quite difficult as *N* increases because the number of false energy minima increases rapidly, particularly after $N = 12$. For this reason alone, we should not expect the Neolithic stone carvers to have discovered the optimum solutions for keeping the knobs apart for larger numbers, but a comparison between what they achieved and the calculations is interesting. A complete analysis would require examination of the objects themselves, which are spread across several museums; here only a partial analysis, based on published illustrations and descriptions, is attempted.[16]

The least number of knobs recorded is three, and there are five examples of this. The Marshall description of the set reads, "Two of the balls are atypical, having rounded projecting knobs making a more or less triangular object which is oval in section. The others have clear cut knobs." It seems likely from this description that all the balls are more or less of planar trigonal symmetry, as expected from the surface Coulomb calculations. Many 4K balls have been illustrated, and all are clearly approximations to the expected tetrahedral symmetry.

However, the execution is of variable quality; the one shown on the Ashmolean Museum website has one of the "three-fold" axes at least 20° away from the expected position. Tetrahedral geometry is difficult to work with since there is no direction from which a carver could check his or her work as it proceeded by "looking for opposites" (Edmonds 1992, 179), as in the next example.

In contrast to the geometrical variations with the tetrahedral 4K, the octahedral 6K balls seem to be more accurately executed, as well as being by far the commonest type. This situation may be due to the fact that a pair of poles and an equator are relatively easy to mark on a sphere by eye. After this step, two perpendicular polar great circles can be added, and there is one beautifully decorated ball, with no deep carving, on which exactly this pattern of three great circles is seen.[17] Such a marking with perpendicular great circles allows the construction of octahedral 6K objects by using the intersections as markers for the centers of the knobs to be carved, but they could also provide a route to the 4K objects by using four of the centers of the spherical triangles as markers.

After the 4K and 6K balls, there are more examples of 7K than of any other.

Electron pair repulsion, and the surface Coulomb calculations, both predict a pentagonal bipyramidal structure, with five knobs in a plane, and most carvers seem to have discovered this structure. Seven of those illustrated show this; particularly good examples are at the Ashmolean Muscum[18] and at the Hunterian Museum.[19] The latter museum also has an example of 7K that has an asymmetric structure with very oval shapes for the knobs[20]; this shaping might be a consequence of the carver having to compensate after missing the symmetric structure.

The 8K objects clearly presented the carvers with some difficulty, since they seem to have been uncertain about the best form to use. Because this object is the only small unit that does not work as a deltahedron, it is hardly surprising that they did not discover the optimum square antiprismatic structure. However, there is an apparently clear path to the cube through the polar circle construction described above, which divides the sphere into eight spherical triangles, as illustrated by the tapes on the 8K object at the left in Figure 1. The fact that so many tetrahedra but so few cubes have been found almost suggests an *avoidance* of the cube structure; more equal numbers would have been expected if the set of Platonic solids had any significance for the carvers.

There is only one 9K object for which an illustration is available,[21] and this object is badly worn. There is a 10K object in the collection of the Ashmolean Museum, which can be seen on their website. A detailed description of this particular object by Critchlow (1979, 147) makes it very clear that the form is essentially a pair of square pyramids on a common four-fold axis in a "staggered" configuration (that is, rotated against each other by 45°). This method creates a deltahedron that has the same symmetry as that calculated (D_{4d} in the Schoenflies convention used by Atiyah and Sutcliffe 2003). There is a similar object in the Hunterian collection.[22]

From the carver's point of view, the form they chose for 12K might well have seemed to be an extension of the 10K structure, a staggered pair of pentagonal pyramids, but the result has the much higher symmetry of the icosahedron, the Platonic structure predicted by the calculations. There are two examples of this in Figure 1, and there is a third in the Dundee museum (Critchlow 1979, 145).

Some of these suggested structural principles outlined still apply to the two 14K objects that have been illustrated. Both are pairs of hexagonal pyramids, and the one in the Ashmolean Museum has the pyramids in a staggered configuration, as in the calculated structure (D_{6d} symmetry).[23] The other example has the pyramids related by a reflection plane,[24] an "eclipsed" geometry. The simple "keep the knobs apart" principle seems to be joined by other influences now, but as noted earlier, even modern calculations of structure become difficult after $N = 12$. For higher numbers, the predicted deltahedral structures become less common among the carved spheres; one possible reason for this phenomenon may be that eclipsed structures are easier to construct. The knobs have all been generated from spheres by grinding away grooves, and a reflection plane allows two or more knobs to be ground together with a single longer groove.

It seems clear that the high symmetries of the objects with relatively small numbers of knobs arise quite naturally from the simple principle of keeping knobs away from each other, or, equivalently, maintaining even spacing. Three of the Platonic solids are generated automatically in this way, and the large numbers found for the tetrahedron and octahedron suggest that the carvers, or their sponsors, found something attractive about these structures, quite possibly connected to what we would call "aesthetic" considerations. However, the gross disparity in numbers between the octahedra and the cubes is a fairly clear indication that neither the concept of regularity, nor any of the other mathematical aspects of the Platonic solids, were understood at this time.

Conclusions

Although there is high-quality craftsmanship in these Neolithic objects, there seems to be little or no evidence for what we would recognize as mathematical ideas behind them. In particular, there is *no* evidence for a prehistoric knowledge of the set of five Platonic solids, and it seems inappropriate to describe the objects as "models" of anything. The conventional historical view that the discovery of the concept of the set of five regular solids was contemporary with Plato can still be taught, and it is unchallenged by the existence of these beautiful objects. Reasons for producing them should be sought in disciplines such as aesthetics and anthropology rather than in mathematics.

Postscript

There is no evidence that the Neolithic sculptors ever made a stone dodecahedron. However, the set of five stone Platonic solids has now been completed, after a delay of five millennia, by the contemporary British sculptor Peter Randall-Page. His "Solid Air II" is shown in Figure 2. The base is one of the pentagonal groups of balls, so one of the

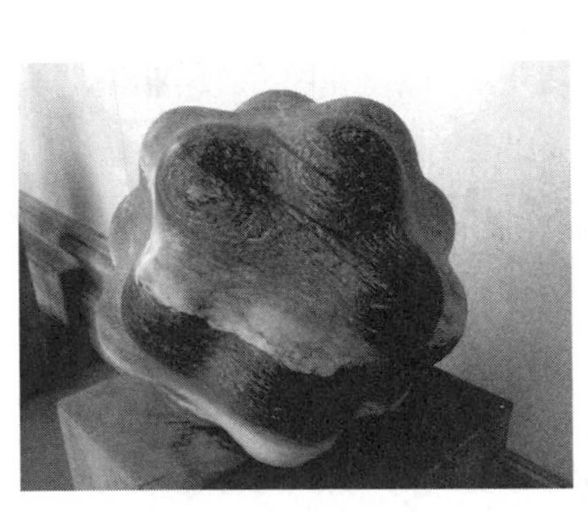

FIGURE 2. "Solid Air II" by Peter Randall-Page. Used by permission of the artist.

five-fold axes of this dodecahedron is vertical. The first photograph is taken close to one of the other five-fold axes, and shows a pentagonal face with five others surrounding it.

Acknowledgments

I thank Graham Challifour for permission to reproduce his photograph (Figure 1) and Alan Saville of The National Museum of Scotland, Tracey Hawkins of the Glasgow Museums, and Sally-Ann Coupar of the Hunterian Museum of Glasgow, for help with various objects in the collections of their museums.

Notes

1. Certainly the tetrahedron, in the form of an early gaming die, is much older. For an entertaining account of this and other aspects of the history of the regular solids, see du Sautoy (2008, 40–58).

2. It can also be argued that the description "most beautiful" suggests that Plato had some idea of what we call symmetry (Lloyd 2010).

3. The original claim was "by a millennium." In fact, given the current dating of the objects, this claim could be extended to "more than two millennia."

4. The claims in Critchlow's book were repeated by Lawlor (1982, 97), whose book includes a better reproduction of the photograph. Figure 1 is taken from Lawlor's version.

5. The National Museum of Scotland, Edinburgh. Items from this collection are referred to by their NMS numbers in subsequent notes. Illustrations can be found on their website: http://nms.scran.ac.uk/search/?PHPSESSID=nnog07544f585b9idc2sftdc13.

6. The 5K objects are atypical, in that these two are heavily decorated with additional carvings, and a third is described as "oval" rather than spherical; these objects are not considered further.

7. There are examples with much higher numbers of knobs, extending to well over 100, but for almost all of these examples, only one, two, or zero are known (three each for 27K and for 50K).

8. NMS 000-180-001-392-C, 000-180-001-369-C, and 000-180-001-328-C.

9. http://www.ashmolean.org/ash/britarch/highlights/ stone-balls.html.

10. NMS 000-180-001-363-C.

11. According to a personal communication from G. Challifour, all the objects in his photograph (Figure 1) were from the same museum.

12. NMS 000-180-001-392-C.

13. NMS 000-180-001-719-C.

14. NMS 000-180-001-368-C.

15. The title of this web page includes the word "hoax." In the correspondence on this page, I have pointed out that there is no evidence to support any such suggestion; see also Hart (1998, Addendum 2009).

16. In addition to those in the NMS (see footnote 5), several useful illustrations can be found in Marshall 1977, and a few in Critchlow 1979; five more are available at the Ashmolean

site (footnote 9). The Hunterian Museum and Art Gallery in Glasgow also has a large collection, but access to the illustrations requires the museum reference number.

17. Marshall 1977, Figure 9:3; the object is AS 16 at the Carnegie Inverurie Museum.

18. Note 9; the curator's drawing shows this structure rather better than the photograph.

19. Two views of this item, GLAHM B.1951.245d, are available at http://tinyurl.com/yhqhlxj.

20. Hunterian Museum GLAHM B.1951.112.

21. Hunterian Museum GLAHM B.1914 349.

22. Hunterian Museum GLAHM B.1951.876.

23. There is another illustration of this phenomenon in Critchlow (1979, 147) with a description that confirms this symmetry.

24. As photographed, the six-fold axis is almost vertical and enters through the left-hand top knob (see Hunterian Museum GLAHM B.1951.245a).

References

Artmann, B., *Euclid–the creation of mathematics*, Springer, 1999.

Atiyah, Michael, *Collected works*, Vol. 6, Oxford University Press, 2004.

Atiyah, Michael, and Sutcliffe, P., "Polyhedra in physics, chemistry and geometry," *Milan Journal of Mathematics*, 71 (2003), 33–58, http://arxiv.org/abs/math-ph/0303071.

Critchlow, Keith, *Time stands still: new light on megalithic science*, Gordon Fraser, 1979.

du Sautoy, Marcus, *Finding moonshine,* Fourth Estate, 2008.

Edmonds, M. R., "Their use is wholly unknown," in *Vessels for the ancestors*, N. Sharples and A. Sheridan (eds.), Edinburgh University Press, 1992.

Erber, T., and Hockney, G. M., "Complex systems: equilibrium configurations of n equal charges on a sphere ($2 < n < 112$)," *Advances in Chemical Physics*, 98 (1997), 495–594.

Hart, G. W., 1998, http://www.georgehart.com/virtual-polyhedra/neolithic.html. Accessed February 2012.

Heath, Thomas, *The thirteen books of Euclid's Elements*, Dover publications, 1956.

Heath, Thomas, *A history of Greek mathematics*, Vol. I, Dover publications, 1981.

Lawlor, R., *Sacred geometry: philosophy and practice*, Thames and Hudson, 1982.

le Bruyn, L., 2009, http://www.neverendingbooks.org/index.php/the-scottish-solids-hoax.html. Accessed February 2012.

Lloyd, D. R., "Symmetry and Beauty in Plato," *Symmetry*, 2 (2010), 455–465.

Marshall, D. N., "Carved stone balls," *Proceedings of the Society of Antiquaries of Scotland*, 108 (1977).

Pekkonen, O., "Atiyah, M., Collected works, vol. 6" (review), *The Mathematical Intelligencer,* 28 (2006), 61–62.

Waterhouse, W. C., "The discovery of the regular solids," *Archive for the History of Exact Sciences*, 9 (1972), 212–221.

Early Modern Mathematical Instruments

JIM BENNETT

For the purpose of reviewing the history of mathematical instruments and the place the subject might command in the history of science, if we take "early modern" to cover the period from the 16th to the 18th century, a first impression may well be one of a change from vigorous development in the 16th century to relatively mundane stability in the 18th. More careful scrutiny suggests that this perception is relative and depends more on our priorities as historians than on the interests of the instrument manufacturers or users. In fact, the role of mathematical instruments and the agenda of their designers and makers were fairly steady over the period, and what changed was how these might be viewed in the broader context of natural philosophy and its instruments. Following the mathematical thread in instrumentation consistently through this time has not been a popular option, even for instrument historians and even though it has a demonstrable, categorical presence. It is all too easy to be diverted by telescopes, microscopes, barometers, air pumps, electrical machines, and the like, once these seductive newcomers arrive on the scene, parading their novel and challenging entanglement with natural knowledge. Yet, as a category, "mathematical instruments" retained its currency and meaning throughout the period, while commercial, professional, and bureaucratic engagement with such objects continued to expand. It remained more common to want to know the time than the barometric pressure, and, despite the growing popularity of orreries, there was still more money in octants.

When historians encounter mathematical instruments, such as astrolabes, sundials, quadrants, surveyors' theodolites, or gunners' sights and rules, it is generally not in the secondary histories where they first learn their trade but, instead, in museums. Coming from book-learned history, with its recent material turn in the history of science, and looking for a material culture from the 16th and 17th centuries can be a dispiriting and perplexing experience. Collections rich in material from the period present uncompromising rows of instruments that are clearly challenging in their technical content but seem obsessed with the wrong kinds of questions. Sundials are plentiful, whereas telescopes and microscopes are fabulously rare. Horary quadrants and gunners' sights present themselves in baffling varieties, but not a single air pump seems to survive.

The dominant instrument culture up to the end of the 17th century characterized itself as "mathematical." By then it incorporated a range of applications of mainly geometrical techniques to an array of what had become, at least in aspects of their practice, *mathematical* arts. Astronomy had set the pace for instrumentation. By the 16th century, instrumentation had long been an integral part of its practice—instruments for the immediate requirement of positional measurement, leading on to those with more particular functionality, such as astrolabes, sundials, and horary quadrants. The armillary sphere (Figure 1) may well have begun as a measuring instrument, but because its arrangement of circles reflected those used by astronomers for registering the heavenly motions, it could also be used for teaching the practice of astronomy and for a limited amount of calculation.

The armillary sphere perfectly indicates the ambiguity of some mathematical instruments, in turn reflecting tensions in the discipline of astronomy itself: Are the circles and motions of the instrument intended primarily to correspond to the heavens or to the geometrical practice of astronomers? Many books in the "sphaera" tradition, popularized in the work of Sacrobosco,[1] can be read equally easily with reference to the sky or to an artificial instrument, the armillary sphere. The emergence of instruments as a prominent instantiation of disciplinary practice may be more profound than has generally been allowed, to be acknowledged alongside instantiation through treatises, terminologies, diagrams, constructions, and rules. An important character of mathematical instruments is that they face more toward disciplinary practice than toward the natural world. That is where they derive their regulation and legitimacy, while the discipline in turn is partly characterized through its engagement with artificial instruments.

In seeking to characterize the role of instruments in disciplinary practice, we can look at how they appear in published treatises.

FIGURE 1. Armillary sphere by Carlo Plato, Rome, 1598. (Museum of the History of Science, Oxford, inventory no. 45453.)

Descriptions are found within more general astronomical works, when some account is needed of the nature of measurement, and an important example of this would be Ptolemy's *Almagest* itself. But even before the first printed edition (1515) of *Almagest*, an independent literature of printed treatises on mathematical instruments was launched in 1513 by Johann Stoeffler with his *Elucidatio fabricae ususque astrolabii.*[2] The formula of "construction and use" was adopted for many of the accounts of particular mathematical instruments that were published throughout the century. Stoeffler first tells his readers how to make an astrolabe, then, in a series of worked examples, how to use it. The treatise presents itself not in the context of an overarching discipline but as an independent manual for a particular instrument.

Through the 16th century, many accounts of individual instruments were published, alongside treatises on sets or selections of instruments, whether closely related in type, such as in Sebastian Münster's books on sundials, or covering as broad a range as possible, as in Giovanni Paolo Gallucci's *Della fabrica et uso di diversi stromenti di astronomia et cosmographia* (1597).[3] This relative detachment from what might be considered the parent or foundational discipline, seen in the publication of books devoted solely to instruments and their use, is worth a moment's thought. Instrument development had a narrative that ran alongside the parent discipline but was not dependent on changes or advances in astronomy. The originality of astronomers and other mathematicians could be exercised through the design and improvement of instruments; a network of instrumental relations could spread among separate disciplines (astronomy, navigation, surveying, architecture, warfare) based on practical techniques deployed in instruments in different fields of practice; a community of makers promoted the development of these links and of instrument design and use as a commercial imperative, while a growing number of practitioners did the same in the hope of professional advantage. In short, instruments had a life that was not entirely dependent on an academic mathematical discipline, and this phenomenon is reflected in the range of 16th century publications.

At the end of the century, the rise of the independent treatise on one or more mathematical instruments reflects back into astronomical measurement itself through the publication of Tycho Brahe's *Astronomiae instauratae mechanica* of 1598.[4] This detailed account of his observatory and its individual instruments, set out one after another in a thorough and comprehensive manner, established an influential precedent for ambitious observatory astronomers; later examples in our period include Johannes Hevelius, John Flamsteed, and Ole Roemer.

Before moving the narrative into the 17th century, and as a narrative thread to lead us there, we might briefly consider the subject of dialing and the division of instruments—sundials and horary quadrants—that characterized the subject. If there is a mismatch between the interests of early modernists and those of their actors with regard to mathematical instruments generally, the disparity is even greater when it comes to dialing. Not only might the problem of finding time from the sun, mostly with portable rather than fixed instruments, be solved in a great variety of imaginative and technically challenging ways, be applicable to different conventions of dividing the day or night and numbering the hours, be made relevant to the whole Earth instead of a single latitude, and so on; but also, the ambitions of such instruments could extend far beyond telling the time "here and now"— or even "there and then." Their functionality could extend outside time telling into other areas of astronomical problem solving.

In the 16th century, dialing was a vibrant, challenging, even competitive astronomical and geometrical discipline, with a large following, many new designs of instruments, and a correspondingly healthy output of publications. It was related closely to the popular contemporary discipline of cosmography—the mathematical, or part-mathematical, discipline dealing with the geometrical relationships between the heavens and the Earth. Leading mathematicians, such as Regiomontanus, Peter Apianus, Gemma Frisius, and Oronce Fine, were centrally involved, seeing dialing as an integral part of their disciplinary practice.[5] Yet dialing is now all but ignored by historians of science. Extending our appreciation to this aspect of contemporary astronomy broadens our grasp of its scope and methodology. Planetary theory is important, of course, but to achieve a rounded account of the discipline we cannot afford to neglect an aspect that was clearly significant for many of its practitioners.

Following the development of dials and quadrants takes us seamlessly into the 17th century and the rise in importance of English mathematicians and makers. The institutional mission of Gresham College in London, with its recognition of the practical mathematical arts and sciences as a subject for the kind of systematic treatment that might underpin a series of weekly lectures, was important for bringing the English into a European movement—one that combined learning, technical innovation, practical application, publication, manufacture, and commerce. The professor of astronomy from 1619 to 1626, Edmund Gunter, had a particular commitment to, and success with, instrumentation, and his eponymous quadrant (Figure 2) established itself as a standard portable astronomical instrument through the 17th and 18th centuries.[6] Characteristic of the sundial's functionality being extended beyond time telling, by restricting his projection of hour lines to a particular latitude, Gunter was able to include other lines, such as the horizon, the ecliptic, and lines of solar azimuth, so that his quadrant could be applied to a range of astronomical calculations. Gunter worked closely with the instrument maker Elias Allen, whose shop near St. Clement's Church[7] in the Strand was a place of resort and exchange for many interested in practical mathematics.

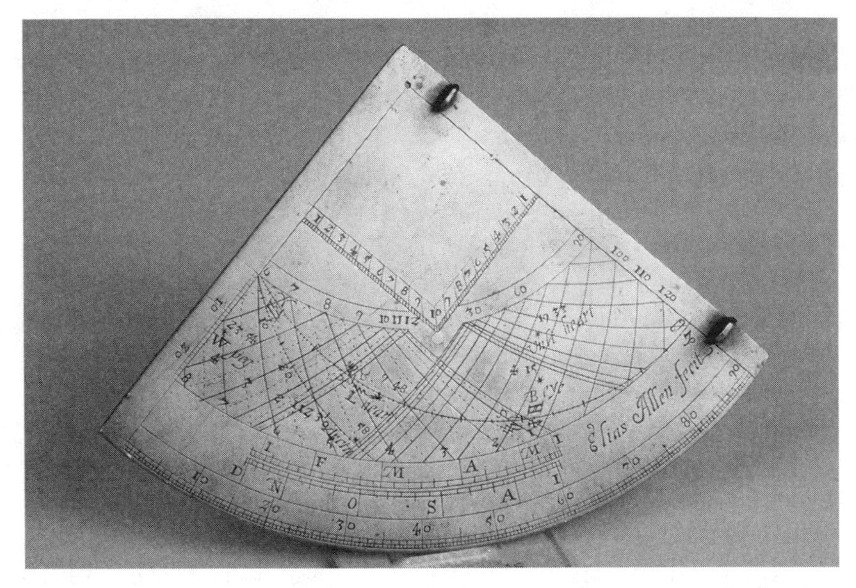

FIGURE 2. Gunter quadrant by Elias Allen, circa 1630. Courtesy of Whipple Museum of the History of Science, Wh. 1764.

Another of Allen's mathematical patrons was William Oughtred, whose design of the universal equinoctial ring dial, like the Gunter quadrant, was popular through the 18th century. Heir to the dialing literature from 16th century continental mathematicians, Oughtred's design was developed from the astronomer's rings of Gemma Frisius. A design of his own was his "horizontal instrument," a dial based on a projection onto the horizon of the sun's daily paths throughout the year, the ecliptic, and the equator; like Gunter's quadrant, it had a range of astronomical functionality. There were fixed and portable versions, and the latter were combined with one of the most important achievements of the new English school of mathematical instrumentation, the delivery of logarithmic calculation as a practical technique by means of an instrument. The close links between publication and instrument manufacture are maintained: The horizontal instrument and the "circles of proportion" (Oughtred's circular logarithmic rule) occupied either side of a portable instrument (that could be bought in the shop of Elias Allen) and appeared together in Oughtred's book of 1632 (published and sold by Allen).[8]

The development of specialized or professionalized calculation around this time is an instructive episode for understanding the nature of mathematical instruments. Imperatives from the world beyond mathematics—in warfare and navigation—were creating opportunities for ambitious mathematicians, adaptable practitioners, and entrepreneurial makers. The kinds of proportional calculations required of the gunner in working out the weight of shot or the appropriate charge of powder were amenable to geometrical handling using similar triangles and could readily be rendered instrumental by a "sector" (or "compass of proportion"), where a pair of scales on the faces of flat rules connected by a compass joint could be opened to different angles, according to the proportionality required. Galileo's "geometrical and military compass" is only the best known of a range of such instruments, where combinations of different pairs of lines might handle not just the direct proportions needed, say, by surveyors drawing plans to scale, but also lines for areas (useful again to surveyors as well as carpenters) and volumes (used by, for example, masons, architects, and gunners).[9]

The sector embodied a very adaptable technique, and, after the mathematics of the Mercator chart for sailing had been explained by another mathematician in the Gresham circle, Edward Wright, Gunter designed a specialized version for managing the kind of trigonometrical calculations required when working with a chart where the scale varied as the secant of the latitude. Gunter went on to devise a rule with lines carrying logarithmic and trigonometric functions so that multiplications and divisions could be carried out by the straightforward addition or subtraction of lengths transferred by means of a pair of dividers. Two such rules (dispensing with the dividers) created the once-familiar logarithmic slide rule. Logarithms had only recently been invented by John Napier, and their rapid application in Gunter's rule and Oughtred's circles indicates the continuing significance of the role played by instruments in contemporary mathematics.

The practical discipline that had introduced such technical novelties as these maintained a strong commercial presence throughout the 17th century, now with the addition of English workshops to those of continental Europe. Indeed, as latecomers to the field, the English (which meant largely London) makers seemed particularly active. It is worth stressing that mathematical instrument makers were specialist craftsmen who formed an identifiable and readily understood trade.

In some continental cities there was a measure of guild regulation, but in London, although the makers had to belong to some London company, it did not matter which, and there was no nominated home for these mathematical workers. So, they are found in the lists of the Grocers, the Drapers, and many others. In the 17th century, the arts of war became somewhat less evident in the literature and surveying somewhat more so; the two fields are linked in the technique of triangulation and the instruments designed to facilitate it. In surveying, these methods are used for drawing maps; in a military context, they are used in range finders.

Mathematical instrument makers did not become involved in the production of telescopes or microscopes. Makers of optical instruments, if they were not astronomers themselves turning their hands to practical work, emerged from among the most skilled and ambitious of the spectacle makers, while mathematical instrument makers continued their independent and customary trade, applying their engraving skills to practical ends in the mathematical arts. They were not concerned with discoveries in the far away or the very small. But there was one area of practical optics that came to impinge on their world—namely, in measuring instruments for astronomy, where telescopic sights were applied to divided instruments to increase the accuracy of the alignment of the index on the distant target. In a superficial sense, this development witnessed a connection between the two separate areas of work: An optical instrument was combined with an astronomical measuring instrument, and the division between optical and mathematical had been breached. But the added telescope was intended simply to improve the accuracy of the measurement; it did not alter the fundamental function of the instrument. The same conjunctions were made in surveying and navigation, with the addition of telescopic sights to designs of theodolite and sextant, but more effectively in the 18th century.

Retailing (if not manufacturing) across the division between mathematical and optical instruments came in a more profound way toward the end of the 17th century; early instances were found in the practices of the London makers Edmund Culpeper and John Rowley. It was here that the relatively loose regulation of the city companies in controlling manufacturing boundaries was a significant advantage over more restrictive regimes. So the 18th century opened with a commercial if not a conceptual link between the two areas of instrumentation, and, as the popularity of experimental natural philosophy grew, a further commercial opportunity arose through the potential for a trade in the apparatus of natural philosophy, such as air pumps, electrical machines, and all the burgeoning content of the "cabinet of physics."

Much of the staple fare of the mathematical instrument maker remained in his repertoire through the 18th century, while again new designs were added. Sundials and quadrants were as popular as ever—before being tempted to add "despite increasing numbers of pocket watches," remember that watches had to be set to time. Though with roots in designs by Hooke and Newton, angular measurement by the principle of reflection now found original and successful applications in octants, sextants, and circles, used in navigation and surveying, while more ambitious theodolites, perhaps adding an altitude arc or circle and a telescope to the azimuth function (Figure 3), made a significant impression on the practice of surveyors. Professional mathematical practitioners such as these came to be associated even more than previously with their instrumentation, while the numbers of mathematical navigators and surveyors continued to grow. An informal professionalization was underway, not regulated by institutes but encouraged through academies, more often private than public, through textbooks, and through the efforts of instrument makers.

There are three general reasons why the trade in mathematical instruments comes back from the margins of vision and into the range of historians of science. One reason arises from the entrepreneurial efforts of a number of makers to deal across all classes of instrument: in mathematics, optics, and natural philosophy. Prominent examples in London would be George Adams and Benjamin Martin—neither primarily identified with mathematical instruments but both including such instruments in their comprehensive range. They were also prominent in the rise of public, commercial natural philosophy and the growing fashion for attending lectures and buying books and instruments. This popularity drew octants and theodolites into the same commercial project as orreries and microscopes.

Second, the leading mathematical instrument makers became responsible for building the major instruments in the growing number of astronomical observatories, whose chief concern was astronomical measurement. It was in the 18th century that it first became normal to turn to commercial makers with commissions for observatory

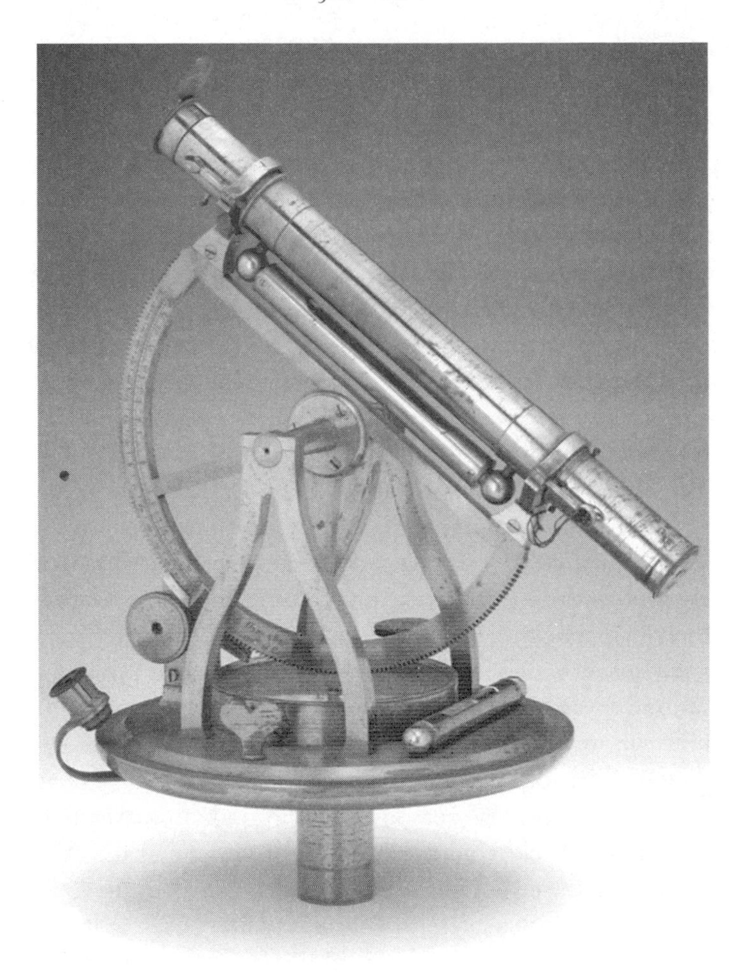

FIGURE 3. Theodolite by George Adams, London, late 18th century. (Museum of the History of Science, Oxford, inventory no. 71425.)

instruments. The recipients of such commissions were, of necessity, the leading mathematical makers, and, for reasons of status if nothing else, they retained their "mathematical" identity. Status was particularly evident in London, where these elite makers were patronized by the Board of Longitude and the Royal Observatory and could become Fellows of the Royal Society and be published in the *Philosophical Transactions*. Toward the end of the century, mathematical instrument makers such as Jesse Ramsden and Edward Troughton came to occupy positions of such respect and influence that they were bound to find their places in the history of the period's science, yet there are sectors signed "Ramsden" and sundials by "Troughton."[10]

Third, the growing intellectual investment in accuracy of observation and measurement in natural investigation could be realized only through the mechanical knowledge and skill that had built up through centuries and now lay in the hands and workshops of mathematical instrument makers.

Although a greater preoccupation with optical and experimental instruments is understandable, historians should not neglect the mathematical. Without understanding that category, and the categorization of instrument making as a whole in the period, we cannot properly appreciate other, more compelling, aspects of the instrument narrative. Mathematical instruments have a much longer history than optical and natural philosophical instruments, and although their established existence contributes to the possibility of instruments with other ambitions, the later types pursue separate developments, in different workshops and areas of practice, fully merging only later in post hoc commercial convenience. It is from this commercial union that the "scientific instrument" emerged. Furthermore, it is the practical mathematical work that engages with the worlds of commerce, bureaucracy, war, and empire, all of which speak to the breadth of the history of science as practiced today. Finally, if we want to bring the likes of Ramsden and Troughton into our mainstream narrative, as we surely must given the importance ascribed to them by their contemporaries, that can be done only with an appreciation of the disciplinary tradition they recognized and through which they understood and defined their own work—that of the mathematical instrument.

Notes

1. Johannes de Sacrobosco wrote his basic textbook on astronomy in c. 1230, and it was known from manuscript and printed versions (variously titled *De sphaera mundi*, or *Tractatus de sphaera*, or simply *De sphaera*) for some four centuries.

2. Johann Stoeffler, *Elucidatio fabricae ususque astrolabii* (Oppenheim, 1513).

3. Sebastian Münster, *Compositio horologiorum* (Basel, 1531); Münster, *Horologiographia* (Basel, 1533); and Giovanni Paolo Gallucci, *Della fabrica et uso di diversi stromenti di astronomia et cosmographia* (Venice,1597).

4. Tycho Brahe, *Astronomiae instauratae mechanica* (Wandsbeck, 1598).

5. Jim Bennett, "Sundials and the Rise and Decline of Cosmography in the Long 16th Century," *Bulletin of the Scientific Instrument Society*, 2009, no. 101, pp. 4–9. More generally, see Hester Higton, *Sundials at Greenwich* (Oxford: Oxford Univ. Press, 2002).

6. Edmund Gunter, *The Description and Use of the Sector, the Crosse-Staffe, and Other Instruments* (London,1624).

7. St. Clement's Church was the name commonly used in the 17th century for the church now known as St. Clement Danes.

8. William Oughtred, *The Circles of Proportion and the Horizontall Instrument* (London, 1632).

9. Filippo Camerota, *Il compasso di Fabrizio Mordente, per la storia del compasso di proporzione* (Florence: Olschki, 2000).

10. Anita McConnell, *Jesse Ramsden (1735–1800): London's Leading Scientific Instrument Maker* (Aldershot: Ashgate, 2007).

A Revolution in Mathematics? What Really Happened a Century Ago and Why It Matters Today

Frank Quinn

The physical sciences all went through "revolutions": wrenching transitions in which methods changed radically and became much more powerful. It is not widely realized, but there was a similar transition in mathematics between about 1890 and 1930. The first section briefly describes the changes that took place and why they qualify as a "revolution," and the second section describes turmoil and resistance to the changes at the time.

The mathematical event was different from those in science, however. In science, most of the older material was wrong and discarded, whereas old mathematics needed precision upgrades but was mostly correct. The sciences were completely transformed while mathematics split, with the core changing profoundly but many applied areas, and mathematical science outside the core, relatively unchanged. The strangest difference is that the scientific revolutions were highly visible, whereas the significance of the mathematical event is essentially unrecognized. The section "Obscurity" explores factors contributing to this situation and suggests historical turning points that might have changed it.

The main point of this article is not that a revolution occurred, but that there are penalties for not being aware of it. First, precollege mathematics education is still based on 19th century methodology, and it seems to me that we will not get satisfactory outcomes until this approach changes [9]. Second, the mathematical community is adapted to the social and intellectual environment of the mid- and late 20th century, and this environment is changing in ways likely to marginalize core mathematics. But core mathematics provides the skeleton that supports the muscles and sinews of science and technology; marginalization will lead to a scientific analogue of osteoporosis. Deliberate management [2] might avoid this, but only if the disease is recognized.

The Revolution

This section describes the changes that took place in 1890–1930, drawbacks, objections, and why the change remains almost invisible. In spite of the resistance, it was incredibly successful. Young mathematicians voted with their feet, and, over the strong objections of some of the old guard, most of the community switched within a few generations.

Contemporary Core Methodology

To a first approximation, the method of science is "find an explanation and test it thoroughly," whereas modern core mathematics is "find an explanation without rule violations." The criteria for validity are radically different: Science depends on comparison with external reality, whereas mathematics is internal.

The conventional wisdom is that mathematics has always depended on error-free logical argument, but this is not completely true. It is quite easy to make mistakes with infinitesimals, infinite series, continuity, differentiability, and so forth, and even possible to get erroneous conclusions about triangles in Euclidean geometry. When intuitive formulations are used, there are no reliable rule-based ways to see that these are wrong, so in practice, ambiguity and mistakes used to be resolved with external criteria, including testing against accepted conclusions, feedback from authorities, and comparison with physical reality. In other words, before the transition, mathematics was to some degree scientific.

The breakthrough was the development of a system of rules and procedures that really worked, in the sense that, if they are followed carefully, then arguments without rule violations give completely reliable conclusions. It became possible, for instance, to see that some intuitively outrageous things are nonetheless true. Weierstrass's nowhere-differentiable function (1872) and Peano's horrifying space-filling curve (1890) were early examples, and we have seen much stranger things since. There is no abstract reason (i.e., apparently no proof) that such a useful system of rules exists and no assurance that we would find it. However, it does exist and, after thousands of years of tinkering and under intense pressure from the sciences for substantial progress, we did find it. Major components of the new methods are the following:

> **Precise definitions:** Old definitions usually described what things are supposed to be and what they mean, and extraction of properties relied to some degree on intuition and physical experience. Modern definitions are completely self-contained, and the only properties that can be ascribed to an object are those that can be rigorously deduced from the definition.
>
> **Logically complete proofs:** Old proofs could include appeals to physical intuition (e.g., about continuity and real numbers), authority (e.g., "Euler did this, so it must be okay"), and casual establishment of alternatives ("these must be all the possibilities because I can't imagine any others"). Modern proofs require each step to be carefully justified.

Definitions that are modern in this sense were developed in the late 1800s. It took a while to learn to use them: to see how to pack wisdom and experience into a list of axioms, how to fine-tune them to optimize their properties, and how to see opportunities where a new definition might organize a body of material. Well-optimized modern definitions have unexpected advantages. They give access to material that is not (as far as we know) reflected in the physical world. A really "good" definition often has logical consequences that are unanticipated or counterintuitive. A great deal of modern mathematics is built on these unexpected bonuses, but they would have been rejected in the old, more scientific approach. Finally, modern definitions are more accessible to new users. Intuitions can be developed by working directly with definitions, and this method is faster and more reliable than trying to contrive a link to physical experience.

Logically complete proofs were developed by Frege and others beginning in the 1880s, and by Hilbert after 1890 and (it seems to me) rounded out by Gödel around 1930. Again it took a while to learn to use these: The "official" description as a sequence of statements obtained by logical operations, and so forth, is cumbersome and opaque, but ways were developed to compress and streamline proofs without losing reliability. It is hard to describe precisely what is acceptable as a modern proof because the key criterion, "without losing reliability," depends heavily on background and experience. It is clearer and perhaps more important what is *not* acceptable: no appeals to authority or physical intuition, no "proof by example," and no leaps of faith, no matter how reasonable they might seem. As with definitions, this approach has unexpected advantages. Trying to fix gaps in first approximations to proofs can lead to conclusions we could not have imagined and would not have dared conjecture. They also make research more accessible: Rank-and-file mathematicians can use the new methods confidently and effectively, whereas success with older methods was mostly limited to the elite.

Drawbacks

As mathematical practice became better adapted to the subject, it lost features that were important to many people.

The new methodology is less accessible to nonusers. Old-style definitions, for instance, usually related things to physical experience, so many people could connect with them in some way. *Users* found these connections dysfunctional, and they can derive *effective* intuition much faster from precise definitions. But modern definitions have to be used to be understood, so they are opaque to nonusers. The drawback here is that nonusers only saw a loss: The old dysfunctionality was invisible, whereas the new opacity is obvious.

The new methodology is less connected to physical reality. For example, nothing in the physical world can be described with complete precision, so completely rule-based reasoning is not appropriate. In fact, the history of science is replete with embarrassing blunders caused by excessively deductive reasoning; see the section "Hilbert's Missed Opportunities" for context and illustrations. Professional practice now accommodates this problem through the use of mathematical models: Mathematics applies to the model but no longer even pretends to say anything about the fit between model and reality. The earlier connection to reality may have been an illusion, but people saw it as a drawback that had to be abandoned. In the other direction, core mathematics no longer accepts externally verified (experimental) results as "known" because this method would bring with it the same limitations on deductive reasoning that are necessary in science. Even the most seemingly minor flaw sooner or later causes proof by contradiction and similar methods to collapse.

In practice, this issue led to a division into "core" mathematics and "mathematical science." For instance, if numerical approximations of fluid flow seem to reproduce experimental observations, then this could be taken as evidence that the approximation scheme converges. This conclusion does not have the certainty of modern proof and cannot be accepted as "known" in the core sense. However, it is a reasonable *scientific* conclusion and appropriate for mathematical *science*. Similarly, the Riemann hypothesis is incredibly well tested. For scientific purposes it is a solid fact, but it is unproved and remains dangerous for core use. Another view of this development is that, as mathematical methods diverged from those of science, mathematics divided into a core branch that separated from physical science in order to exploit these methods and a mathematical science branch that accepted the limitations in order to remain connected. The drawback here is that the new power in the core and the support it gives to applied areas are invisible to outsiders, whereas the separation from science is obvious. People wonder: Is core mathematics a pointless academic exercise and mathematical science the real thing?

Opposition

Henri Poincaré was the most visible and articulate opponent of the new methods; cf. [6]. He felt that Dedekind's derivation of the real numbers from the integers was a particularly grievous conceptual error because it damaged connections to reality and intuitive understanding of continuity. Some of the arguments were quite heated; the graphic novel *Logicomix* [1] dramatically illustrates the turmoil (though it muddles the issues a bit). Scholarly works [3] are more dignified but give the same picture.

As the transition progressed, the arguments became more heated but more confined. At the beginning, traditionalists were deeply offended but not threatened. But because modern methods lack external checks, they depend heavily on fully reliable inputs. Older material was filtered to support this, and as the transition gained momentum, some old theorems were reclassified as "unproved," some methods became unacceptable for publication, and quite a few ways of looking at things were rejected as dangerously imprecise. Understandably, many eminent late 19th century mathematicians were outraged by these reassessments. These battles were fought by proxy, however. For instance, Poincaré's monumental development of the theory of manifolds was quite intuitive, and we now know that some of his basic intuitions were wrong. But, in the early 20th century, only a fool would have openly criticized Poincaré, and he could not respond to implicit reproaches. As a result, the arguments usually concerned abstractions, such as "creativity" and "understanding," often in the context of education.

On a more general level, scientific concerns about the new methods were reasonable. The crucial importance of external reality checks in physics had been a hard-won lesson, and analogous revolutions in biology and chemistry were still in progress (Darwin's *Origin of Species* appeared in 1859, and Mendeleev's periodic table in 1869). How could mathematical use of the discredited "pure reason" approach possibly be a good thing?

Most of the various schools of philosophy were, and remain, unconvinced by the new methods. Philosophers controlled words such as "reality," "knowledge," "infinite," "meaning," "truth," and even "number," and these were interpreted in ways unfriendly to the new mathematics. For example, if a mathematical idea is not clearly manifested in the physical world, how can it be "real"? And if it is not real, how can it have "meaning," and how can it make sense to claim to "know" something about it? In practice, mathematicians do find that their world has meaning and at least a psychological reality. If philosophy were a science, then this would qualify as a challenge for a better interpretation of "real." But philosophy is not a science. The arguments are plagued by ambiguity and cultural and linguistic biases. "Validation" is mostly a matter of conviction and belief, not functionality, so there are few mechanisms to correct or even expose the flaws. Thus, rather than refine the meaning of "reality" to accommodate what people actually do, philosophers split into Platonists and non-Platonists, depending on whether they believed that mathematics fit their own interpretation. The Platonic view is hard to defend because mathematics honestly does not fit the usual meanings of "real" (see the confusion in Linnebo's overview [5]). The non-Platonic view is essentially that mathematicians are deluded. Neither view is useful for mathematics. To make real progress, mathematics had to break with philosophy and, as usual in a divorce, there are bad feelings on both sides.[1]

The precollege-education community was, and remains, antagonistic to the new methodology. One reason is that traditional mathematicians, most notably Felix Klein, were extremely influential in early 20th century educational reform. Klein founded the International Commission on Mathematical Instruction (ICMI) [4], the education arm of the International Mathematical Union. His 1908 book *Elementary Mathematics from an Advanced Viewpoint* was a virtuoso example of 19th century methods and did a lot to cement their place in education. The "Klein project" [4] is a contemporary international effort to update the *topics* in Klein's book but has no plan to update the *methodology.*[2] In brief, traditionalists lost the battle in the professional community but won in education. The failure of "new math" in the 1960s and '70s is taken as further confirmation that modern mathematics is unsuitable for children. This attempt was hardly a fair test of the methodology because it was poorly conceived, and many traditionalists were determined that it would succeed only over their dead bodies. However, the experience reinforced preexisting antagonism, and opposition is now a deeply embedded article of faith.

Many scientists and engineers depend on mathematics, but its reliability makes it transparent rather than appreciated, and they often dismiss core mathematics as meaningless formalism, obsessive-compulsive about details. This cultural attitude reflects feelings of power in their domains and world views that include little else, but it is encouraged by the opposition in elementary education and philosophy.

In fact, hostility to mathematics is endemic in our culture. Imagine a conversation:

A: What do you do?
B: I am a ___.
A: Oh, I hate that.

Ideally, this response would be limited to such occupations as "serial killer," "child pornographer," and maybe "politician," but "mathematician" seems to work. It is common enough that many of us are reluctant to identify ourselves as mathematicians. Paul Halmos is said to have told outsiders that he was in "roofing and siding"!

Obscurity

Like most people with some exposure to the history of mathematics, I knew about the "foundational crisis" that occurred roughly a century ago. However, my first inkling that something genuinely revolutionary happened came at an international conference on proofs in mathematics education.[3] Sophisticated educators described proofs in ways that I did not recognize, while my description [8], based on an analysis of modern practice [7], was alien to them. The picture that emerged after a great deal of reading and study is that these educators were basing their ideas on insightful analysis of professional practice up through the 19th century. They were not misunderstanding modern mathematics but correctly understanding premodern mathematics. The disconnect stems from a change in mathematics itself, a change of which they were unaware.

No one is unaware of the scientific revolutions. The first subsection suggests that high-profile publicity had a lot to do with this, and obscurity of the mathematical transition is in a sense a public relations failure. To make this problem more concrete, the second section describes some public relations opportunities that Hilbert had but did not use.

Proxies and Belief

Scientific revolutions were methodological, but it was conclusions that attracted attention. The Copernican revolution, for instance, is known for the then-controversial conclusion that the Earth orbits the Sun, and the Darwinian revolution in biology is known for controversial conclusions about human origins. In both cases, the real advances were methodologies effective enough to make alternative conclusions untenable, but methodology is complex and technical. High-profile conclusions served as public proxies for the methodology.

This proxy picture suggests several difficulties for mathematics. First, mathematical conclusions are not exciting enough to provide highly visible proxies. Second, conclusions used to promote mathematics are almost always applications to science, medicine, and engineering. They are proxies for *mathematical science* and have raised visibility of these areas, not the core. For the core, these efforts to use proxies may have actually backfired. Finally, when core results such as the Fermat conjecture or the Poincaré conjecture are described, it is—of necessity—in heuristic terms that are compatible with 19th century viewpoints. The descriptions hide the crucial role of modern methodology, so they are not proxies for it. We will see that there are metamathematical conclusions that at one time might have served as proxies for modern methods, but they were not used.

The science examples also suggest a problem with belief. Users adopt more effective methods, but nonusers often reject things that they do not like (e.g., evolution), regardless of benefits to the technical community. Core methods, such as completely precise definitions (via axioms) and careful logical arguments, are well known, but many educators, philosophers, physicists, engineers, and applied mathematicians reject them as not really necessary. There are cases in which physical science has been unable to overcome rejection based on dislike, so even a clear case for modern mathematics may not succeed.

Hilbert's Missed Opportunities

David Hilbert was the strongest and most highly visible proponent of the new methods during the transition, and as such he was frequently involved in controversies. I describe several situations in which Hilbert might have reframed debates and provided metamathematical proxies that could have led to a much clearer view today. The historical context is used to make the discussion more concrete, not to reproach Hilbert. After all, these opportunities are still just barely visible, even with a century of hindsight. The first controversy occurred early in Hilbert's career and concerned his vigorous use of the "law of the excluded middle" (proof by contradiction). His response to the objections was that denying mathematicians use of this principle was "like denying boxers the use of their fists": true, but not a clear claim or challenge.

If he had said the following, it would have caused an uproar:

> Excluded-middle arguments are unreliable in many areas of knowledge, but absolutely essential in mathematics. Indeed we might *define* mathematics as the domain in which excluded-middle arguments are valid. Instead of debating whether or not it is true, we should investigate the constraints it imposes on our subject.

At the time, mathematics was generally seen as an abstraction of physical reality, and it would have been outrageous to suggest that a logical technique should have higher priority in shaping the field. But in fact nothing physical can be described precisely enough to make excluded-middle arguments reliable, and this problem, as much as anything, drove the division of mathematics. In applied areas, these arguments continued to be tempered by wisdom and experience. In the core, the link to reality became indirect, with modeling as an intermediate, primarily to provide a safe environment for excluded-middle arguments.

Such a statement would have redirected the debate by making successful use of excluded-middle arguments a proxy for core methods. It would also have enabled the issue to be settled in a coherent way. As it was, this issue was a constant pain for Hilbert; Brouwer's Intuitionist school kept it alive into the 1930s; and it died out more from lack of interest than any clear resolution.

Next, Hilbert's axiomatic formulation of geometry in 1899 precisely specified how points, lines, and so forth *interacted*, rather than specifying what they "were" and extracting the interactions from physical intuition. Hilbert himself pointed out that this scheme disconnected mathematics from reality because one could interpret "point" as "chair" and the axioms would remain valid. Again this provoked objections. He might have pointed out that it is a powerful advantage to be able to use a single mathematical construct to model many physical situations. This explanation would have made the disconnect a proxy for mathematics as an independent domain. Widespread acceptance of explicit modeling would then have carried with it acceptance of mathematical independence. As it happened, modeling became widespread in the professional community without being seen as having any such significance.

Hilbert's famous 1900 problems were powerful technical challenges that did a lot to drive development of infinite-precision methods. However, the few that were actually seen as proxies for new ways of thinking (e.g., the second, on consistency of arithmetic) did not fare well, and the changes that the problems helped drive remained mostly invisible.

Another debate concerned the use of axiomatic definitions and detailed logical arguments. These ideas provoked strong objections about lack of reality and meaning, artificial rigidity, and content-free formal manipulation. Hilbert might have replied:

> Axiomatic definitions can be artificial and useless, but they can also encapsulate years, if not centuries, of difficult experience, and newcomers can extract reliable and effective intuitions from them. Similarly, fully detailed arguments can be formal and content-free, but fully confronting all details usually deepens understanding and often leads to new ideas. Fully detailed arguments also give fully reliable conclusions, and full reliability is essential for successful use of the powerful but fragile excluded-middle method.

This statement would have acknowledged the dangers of formality but established reliability as a proxy for high-precision methodology and implicitly staked a claim to a nonphysical sort of meaning. Instead, Hilbert accepted the slanders by saying "mathematics is a game played according to certain rules with meaningless marks on paper." Hilbert also suggested that these mathematical methods might be prototypes for similar developments in other sciences. Such things were in vogue at the time. Arthur Conan Doyle, for instance, set his enormously popular Sherlock Holmes stories in a world where excluded-middle logic actually worked:

> . . . when you have eliminated the impossible, whatever remains, however improbable, must be the truth . . .
>
> —*The Sign of the Four,* ch. 6 (1890)

It was probably not widely known that this sort of logic led Doyle himself to a strong and expensive belief in fairies. Blondlot's "*N*-ray" debacle in France around 1904 was not yet seen as a cautionary tale. Since then there have been quite a few embarrassing failures caused by excessively deductive reasoning in science. In the "cold fusion" episode in 1989, for instance, electrochemists Fleischmann and Pons observed excess energy in some of their experiments. After eliminating electrochemical explanations, they deduced that the only alternative they could imagine—hydrogen fusion in the electrodes—must be the truth. This is a standard move in mathematics and in Doyle's fiction but bad science because there is no way to ensure that all alternatives have been imagined. Good *scientific* practice would have required them to test the fusion deduction, for instance, by looking for the radiation that would have been a byproduct of fusion. Not seeing radiation would have turned them back to interesting electrochemistry. Presumably they had stumbled on a previously unimagined way to make a battery, and it was releasing energy accumulated during earlier experiments. But their reliance on the power of deduction led instead to crashing ends to their careers and reputations.

The modern view is that Hilbert's proposal—that mathematical deduction might be a general prototype for science—is a failure. His linkage ended up casting doubt on mathematical developments instead of justifying them. Meanwhile, high reliability has been achieved in mathematics without drawing attention or having significance attached to it. The axiomatic-definition approach also made mathematics more accessible. A century ago, original research was possible only for the elite. Today it is accessible enough that publication is required for promotion at even modest institutions, and an original contribution can be required for a Ph.D. Again this is a profound change that had no significance attached to it.

The final missed opportunity concerns disagreements about knowledge, meaning, and "true." By 1920, the search for secure foundations had bogged down in obscure philosophical arguments. Hilbert had proposed a precise technical meaning for "true," namely, "provable from axioms that themselves could be shown to be consistent." But 10 years later, Gödel showed that in the usual formulation of arithmetic there are statements that are impossible to contradict but not provable in Hilbert's sense. In particular, consistency of the system could not be proved within the system. This was seen as a refutation of Hilbert's proposal. Ironically, it had the same practical consequences because it established "impossible to contradict" as the precise mathematical meaning of "true." Hilbert might have been explicit about deeper goals, for instance:

> Mathematics needs a precise definition of "true" that is internal and accessible to mathematical verification, and in particular unconstrained by philosophy or imagined connections to physical reality. We can worry about what such a definition "means" after it has been developed and shown to be successful in actual practice.

In this light, Gödel's work would have been seen as a successful modification rather than a refutation.[4] Since that time, a precise internal meaning for "true" has been enormously liberating for professional work, but its benefits have gone unnoticed.

Summary

The mathematical transition had such a low profile that no one understood its significance. Felix Klein was still denouncing the new methods in the 1920s, and because his views were not only unrefuted but almost unchallenged, outsiders accepted them as fact. Historians, educators, and philosophers went forward largely unaffected, propelled by the momentum of 3,000 years and rebuffed by the technical complexity of modern practice.

Strangely, mathematicians are also unaware that their field changed so profoundly. Newcomers found philosophical arguments incomprehensible and irrelevant, and philosophy went from a respectable pursuit to an object of ridicule and evidence of senility in just a few decades. However, this situation replaced bad understanding with no understanding at all. Mathematicians have joined fish and birds in doing something very well without any idea how!

The Core at Risk?

For most of the 20th century, mathematics was mainly supported in the higher educational system. Core mathematicians dominated this system, so mathematics had a secure niche that did not depend on understanding what it was about. However, this niche is eroding, and the security is gone.

A large-scale problem is that resource constraints are eroding the ability of the higher education system to support basic research. There is pressure to increase instructional productivity by replacing researchers with teaching faculty at half the cost. Mathematics departments with large service loads are particularly vulnerable. There is also pressure to increase research productivity, with consequences discussed below.

There is a problem with external research funding. In the United States, external support for core mathematics comes almost exclusively from the National Science Foundation. A desire to have something to show for the money has led the NSF to want "wider impacts," and the use of applications as proxies to promote mathematics has led to "applications" being the default interpretation of "wider impacts." The result is a steady shift of funding toward applied areas (and education; see below). Because external funding is a major indicator of productivity, a decline in NSF support for core activity has contributed to the decline in core activity in academic departments.

Yet another problem comes from changes in applied mathematics. Up through the late 20th century, applied mathematicians were trained in mainstream graduate programs and had foundations in modern methods and values. Today many are several generations removed from these core mathematical foundations. Many are scientists rather than mathematicians in the modern sense, and some are actually hostile to core methodology. At the same time, demand from science and engineering and pressure for more highly visible research have caused many academic departments to shift toward applied areas. The result is culturally divided departments in which core mathematics is increasingly at a disadvantage.

The final problem concerns the disconnect between school mathematics and higher education. School mathematics is still firmly located in the 19th century, so student success rates in modern courses have been low. There is a great deal of pressure to improve this situation, but recent changes, such as the use of calculators and emphasis on vague understanding over skills, have actually worsened the disconnect. Something has to change. Ideally, school mathematics could be brought into the 20th century. Unfortunately the K–12 education community is better organized, more coherent, and far more powerful politically. External funding agencies are committed to the K–12 position. At the NSF this situation means that funds have shifted from research to educational programs that are actually hostile to the research methodology. It seems possible that the K–12 college articulation will be "improved" by forcing higher education to revert to 19th century models.

The point in all these examples is that the nature of modern core mathematics must be much better understood to even see the problems. And if the problems are not recognized and addressed quickly, then—in the United States anyway—core mathematics may well be marginalized, and the mathematical Golden Age that began in the 20th century will end in the 21st.

The key question is, "Why would marginalization of the core be a problem, if one is not particularly interested in the subject itself?" In fact, core mathematics provides a rigid skeleton that supports the muscles of science, engineering, and applied mathematics. It is relatively invisible because it cannot interact directly with the outside world; it grows slowly; and it would not cause immediate problems if it stopped growing. Premodern mathematics and contemporary mathematical science, on the other hand, are more like exoskeletons: in direct contact with reality but putting strong constraints on size and power. The long-term consequence of mathematical osteoporosis is that science would have to go back to being a bug!

Solutions for Education?

The point briefly addressed here[5] is that modern methods were adopted because they are much more effective at advanced levels. If the reasons for their success are clearly understood, then some of these methods might be adaptable to elementary levels. This is the meaning of "brought into the 20th century" in the discussion above, and at the very least it would improve K–12 and college articulation. But it might do far more.

To be specific, consider fractions. Currently these are introduced in the old-fashioned way, through connections with physical experience. This method is philosophically attractive and "easy" but follows the historical pattern (see the discussion in "Drawbacks") of being dysfunctional for most students. If we want students to be able to actually *use* fractions, then core experience points a way: Use a precise definition that looks obscure at first but that can be internalized by working with it and that is far more effective once it is learned. Such an approach is suggested in [9] and elaborated in some of the essays in [10]. Similarly, in [8] I explain how a careful understanding of the nature of modern proofs might improve success even with arithmetic. (These are detailed and specific illustrations but are given as starting points rather than "classroom ready").

The big question here is this: Can any version of these approaches be used by real children? Children are attracted to rule-based reasoning (think *games*), and rich applications and success downstream should more than compensate for initial obscurity. I suspect that it is a bigger challenge for educators to think this way than it would be for children. The starting point would be to acknowledge the significance of the mathematical revolution a century ago and to see the new methods—properly understood—as profoundly rich resources rather than alien threats.

Notes

1. There are exceptions, but I wonder whether some ostensible good feelings might not be instances of another thing seen in divorces: One partner remains in love with a fantasy assembled from the good times they had together. See the "Other Views" section in [7] for instances.

2. For detailed explanation, see the essay "Updating 'Klein's Elementary Mathematics from an Advanced Viewpoint': Content only, or the viewpoint as well?" in [10].

3. ICMI Study 19, Taipei, May 2009.

4. It is doubtful that either Hilbert or Gödel would have accepted this formulation. Both felt that the core axioms of mathematics should be "concrete intuitions," an extramathematical criterion. Their interpretations of "finitistic" were also less well defined and less internal than those used today (Tait [11]). In these ways, Hilbert and Gödel were still not fully modern.

5. And at great length in [9] and [10].

References

[1] A. Doxiadis, C. Papadimitriou, A. Papadatos, and A. Didonna, *Logicomix: An Epic Search for Truth*, Bloomsbury, 2009.

[2] Tom Everhart, Mark Green, and Scott Weidman, Math Sciences 2025, *Notices of the AMS* **58** (July 2011), 765.

[3] Jeremy Gray, *Plato's Ghost: The Modernist Transformation of Mathematics*, Princeton University Press, 2008.

[4] International Commission on Mathematical Instruction, http://mathunion.org/icmi, and ICMI Klein Project, http://www.kleinproject.org/.

[5] Oystein Linnebo, Platonism in the Philosophy of Mathematics, in *Stanford Encyclopedia of Philosophy*, (E. Zalta, ed.) 2011, http://plato.stanford.edu/archives/fall2011/entries/Platonism-mathematics.

[6] Henri Poincaré, *Science et Méthode,* E. Flammarion Pub., Paris, 1908.

[7] Frank Quinn, *A qualitative description of contemporary mathematics*, current drafts available at http://www.math.vt.edu/people/quinn/history_nature/index.html.

[8] ———, *Contemporary proofs for mathematics education,* In *Proof and Proving in Mathematics Education*, edited by Gila Hanna and Michael de Villiers. Dordrecht, Germany: Springer Verlag, 2012.

[9] ———, A science-of-learning approach to mathematics education, *Notices of the AMS* **58** (2011), 1264–1275.

[10] ———, *Essays on mathematical learning,* preprints; current drafts at http://www.math.vt.edu/people/quinn/education/index.html.

[11] W. W. Tait, Gödel on intuition and on Hilbert's finitism, in *Kurt Gödel: Essays for His Centennial* (K. Gödel, S. Feferman, C. Parsons, S. G. Simpson, eds.), Cambridge Univ. Press, 2010, pp. 88–108.

Errors of Probability in Historical Context

Prakash Gorroochurn

1. Introduction

This article outlines some of the mistakes made in the calculus of probability, especially when the discipline was being developed. Such is the character of the doctrine of chances that simple-looking problems can deceive even the sharpest minds. In his celebrated *Essai Philosophique sur les Probabilités* (Laplace 1814, p. 273), the eminent French mathematician Pierre-Simon Laplace (1749–1827) said,

> . . . the theory of probabilities is at bottom only common sense reduced to calculus.

There is no doubt that Laplace was right, but the fact remains that blunders and fallacies persist even today in the field of probability, often when "common sense" is applied to problems.

The errors I describe here can be broken down into three main categories: (i) use of "reasoning on the mean" (ROTM), (ii) incorrect enumeration of sample points, and (iii) confusion regarding the use of statistical independence.

2. Use of "Reasoning on the Mean" (ROTM)

In the history of probability, the physician and mathematician Gerolamo Cardano (1501–1575) was among the first to attempt a systematic study of the calculus of probabilities. Like those of his contemporaries, Cardano's studies were primarily driven by games of chance. Concerning his 25 years of gambling, he famously said in his autobiography (Cardano 1935, p. 146),

> . . . and I do not mean to say only from time to time during those years, but I am ashamed to say it, everyday.

Cardano's works on probability were published posthumously in the famous 15-page *Liber de Ludo Aleae,*[1] consisting of 32 small chapters (Cardano 1564). Cardano was undoubtedly a great mathematician of his time but stumbled on several questions, and this one in particular: "How many throws of a fair die do we need in order to have a fair chance of at least one six?" In this case, he thought the number of throws should be three.[2] In Chapter 9 of his book, Cardano says of a die:

> One-half of the total number of faces always represents equality[3]; thus the chances are equal that a given point will turn up in three throws . . .

Cardano's mistake stems from a prevalent general confusion between the concepts of probability and expectation. Let us dig deeper into Cardano's reasoning. In the *De Ludo Aleae*, Cardano frequently makes use of an erroneous principle, which Ore called a "reasoning on the mean" (ROTM) (Ore 1953, p. 150; Williams 2005), to deal with various probability problems. According to the ROTM, if an event has a probability p in one trial of an experiment, then in n trials the event will occur np times on average, which is then wrongly taken to represent the *probability* that the event will occur in n trials. In our case, we have $p = 1/6$ so that, with $n = 3$ throws, the event "at least a six" is wrongly taken to occur an average $np = 3(1/6) = 1/2$ of the time. But if X is the number of sixes in three throws, then $X \sim B(3,1/6)$, the probability of one six in three throws is 0.347, and the probability of at least one six is 0.421. On the other hand, the expected value of X is 0.5. Thus, although the expected number of sixes in three throws is 1/2, neither the probability of one six or at least one six is 1/2.

We now move to about a century later when the Chevalier de Méré[4] (1607–1684) used the *Old Gambler's Rule,* leading to fallacious results. As we shall see, the Old Gambler's Rule is an offshoot of ROTM. The Chevalier de Méré had been winning consistently by betting even money that a six would come up at least once in four rolls with a single die. However, he had now been losing on a new bet, when in 1654 he met his friend, the amateur mathematician Pierre de Carcavi (1600–1684). De Méré had thought that the odds were favorable on betting that he could throw at least one *sonnez* (i.e., double six) with 24 throws of a pair of dice. However, his own experiences indicated that 25 throws were required.[5] Unable to resolve the issue, the two men consulted their mutual friend, the great mathematician, physicist, and philosopher Blaise Pascal (1623–1662).[6] Pascal himself had previously been interested in the games of chance (Groothuis 2003, p. 10). Pascal must have been intrigued by this problem and, through the intermediary of Carcavi,[7] contacted the eminent mathematician, Pierre de Fermat (1601–1665),[8] who was a lawyer in Toulouse. In a letter Pascal addressed to Fermat, dated July 29, 1654, Pascal says (Smith 1929, p. 552),

> He [De Méré] tells me that he has found an error in the numbers for this reason:
>
> If one undertakes to throw a six with a die, the advantage of undertaking to do it in 4 is as 671 is to 625.
>
> If one undertakes to throw double sixes with two dice the disadvantage of the undertaking is 24.
>
> But nonetheless, 24 is to 36 (which is the number of faces of two dice) as 4 is to 6 (which is the number of faces of one die).
>
> This is what was his great scandal which made him say haughtily that the theorems were not consistent and that arithmetic was demented. But you can easily see the reason by the principles which you have.

De Méré was thus distressed that his observations were in contradiction with his mathematical calculations. His erroneous mathematical reasoning was based on the erroneous Old Gambler's Rule (Weaver 1982, p. 47), which uses the concept of the *critical value* of a game. The critical value C of a game is the smallest number of plays such that the probability the gambler will win at least one play is 1/2 or more. Let us now explain how the Old Gambler's Rule is derived. Recall Cardano's "reasoning on the mean" (ROTM): If a gambler has a probability p of winning one play of a game, then in n plays the gambler will win an average of np times, which is then wrongly equated to the *probability* of winning in n plays. Then, by setting the latter probability to be half, we have

$$C \times p = \frac{1}{2}$$

Moreover, given a first game with (p_1, C_1), then a second game which has probability of winning p_2 in each play must have critical value C_2, where

$$C_1 p_1 = C_2 p_2 \quad \text{or} \quad C_2 \frac{C_1 p_1}{p_2} \quad \text{(Old Gambler's Rule)} \tag{1}$$

That is, the Old Gambler's Rule states that the critical values of two games are in inverse proportion as their respective probabilities of winning. Using $C_1 = 4$, $p_1 = 1/6$, and $p_2 = 1/36$, we get $C_2 = 24$. However, with 24 throws, the probability of at least one double six is 0.491, which is less than 1/2. So $C_2 = 24$ cannot be a critical value (the correct critical value is shown below to be 25), and the Old Gambler's Rule cannot be correct. It was thus the belief in the validity of the Old Gambler's Rule that made de Méré wrongly think that, with 24 throws, he should have had a probability of 1/2 for at least one double six.

Let us see how the erroneous Old Gambler's Rule should be corrected. By definition, $C_1 = [x_1]$, the smallest integer greater or equal to x_1, such that $(1 - p_1)^{x_1} = 0.5$, that is, $x_1 = \ln(0.5)/\ln(1 - p_1)$. With obvious notation, for the second game, $C_2 = [x_2]$, where $x_2 = \ln(0.5)/\ln(1 - p_2)$. Thus the true relationship should be

$$x_2 = \frac{x_1 \ln(1 - p_1)}{\ln(1 - p_2)} \tag{2}$$

We see that Equations (1) and (2) are quite different from each other. Even if p_1 and p_2 were very small, so that $\ln(1 - p_1) \approx -p_1$ and $\ln(1 - p_2) \approx -p_2$, we would get $x_2 = x_1 p_1/p_2$ approximately. This is still different from Equation (1) because the latter uses the integers C_1 and C_2, instead of the real numbers x_1 and x_2.

The Old Gambler's Rule was later investigated by the French mathematician Abraham de Moivre (1667–1754), who was a close friend to Isaac Newton. Thus, in the *Doctrine of Chances* (de Moivre 1718, p. 14), Problem V, we read,

> To find in how many Trials an Event will Probably Happen or how many Trials will be required to make it indifferent to lay on its Happening or Failing; supposing that *a* is the number of Chances for its Happening in any one Trial, and *b* the number of chances for its Failing.

TABLE 1.
Critical values obtained using the Old Gambling Rule, de Moivre's Gambling Rule, and the exact formula for different values of p, the probability of the event of interest

Value of p	Critical value C using the Old Gambling Rule $C = C_1p_1/p$ (assuming $C_1 = 4$ for $p_1 = 1/6$)	Critical value C using de Moivre's Gambling Rule $C = [0.693/p]$	Critical value C using the exact formula $C = [-\ln(2)/\ln(1-p)]$
1/216	144	150	150
1/36	24	25	25
1/6	4	5	4
1/4	3	3	3
1/2	2	2	1

De Moivre solves $(1-p)^x = 1/2$ and obtains $x = -\ln(2)/\ln(1-p)$. For small p,

$$x \approx \frac{0.693}{p} \quad \text{(De Moivre's Gambling Rule)} \tag{3}$$

Let us see if we obtain the correct answer when we apply de Moivre's Gambling Rule for the two-dice problem. Using $x \approx 0.693/p$ with $p = 1/36$ gives $x \approx 24.95$, and we obtain the correct critical value $C = 25$. The formula works only because p is small enough and is valid only for such cases.[9] The other formula that could be used, and that is valid for *all* values of p, is $x = -\ln(2)/\ln(1-p)$. For the two-dice problem, this exact formula gives $x = -\ln(2)/\ln(35/36) = 24.60$, so that $C = 25$. Table 1 compares critical values obtained using the Old Gambler's Rule, de Moivre's Gambling Rule, and the exact formula.

3. Incorrect Enumeration of Sample Points

The Problem of Points[10] was another problem about which de Méré asked Pascal in 1654 and was the question that really launched the theory of probability in the hands of Pascal and Fermat. It goes as follows: "Two players A and B play a fair game such that the player who wins a total of 6 rounds first wins a prize. Suppose the game unexpectedly stops when A has won a total of 5 rounds and B has won a total of 3 rounds. How should the prize be divided between A and B?" To solve the Problem of Points, we need to determine how likely A and B are to win the prize if they had continued the game, based on the number of rounds they have already won. The relative probabilities of A and B winning thus determine the division of the prize. Player A is one round short, and player B three rounds short, of winning the prize. The maximum number of hypothetical remaining rounds is $(1 + 3) - 1 = 3$, each of which could be equally won by A or B. The sample space for the game is

$$\Omega = \{A_1,\ B_1A_2,\ B_1B_2A_3,\ B_1B_2B_3\} \tag{4}$$

Here B_1A_2, for example, denotes the event that B would win the first remaining round and A would win the second (and then the game would have to stop since A is only one round short). However, the four sample points in Ω are not equally likely.

For example, event A_1 occurs if any one of the following four equally likely events occurs: $A_1A_2A_3$, $A_1A_2B_3$, $A_1B_2A_3$, and $A_1B_2B_3$. In terms of equally likely sample points, the sample space is thus

$$\Omega = \{A_1A_2A_3,\ A_1A_2B_3,\ A_1B_2A_3,\ A_1B_2B_3,\ B_1A_2A_3,\ B_1A_2B_3,\ B_1B_2A_3,\ B_1B_2B_3\} \tag{5}$$

There are in all eight equally likely outcomes, only one of which $(B_1B_2B_3)$ results in B hypothetically winning the game. Player A thus has a probability of 7/8 of winning. The prize should therefore be divided between A and B in the ratio 7:1.

The Problem of Points had already been known hundreds of years before the times of these mathematicians.[11] It had appeared in Italian manuscripts as early as 1380 (Burton 2006, p. 445). However, it first came in print in Fra Luca Pacioli's *Summa de Arithmetica, Geometrica, Proportioni, et Proportionalita* (1494).[12] Pacioli's incorrect answer was that the prize should be divided in the same ratio as the total number of games the players had won. Thus, for our problem, the ratio is 5:3. A simple counterexample shows why Pacioli's reasoning cannot be correct. Suppose players A and B need to win 100 rounds to win a game, and when they stop Player A has won one round and Player B has won none. Then Pacioli's rule would give the whole prize to A even though she is a single round ahead of B and would have needed to win 99 more rounds had the game continued![13]

Cardano had also considered the Problem of Points in the *Practica arithmetice* (Cardano 1539). His major insight was that the division of stakes should depend on how many rounds each player *had yet to win*, not on how many rounds they had *already won*. However, in spite of this, Cardano was unable to give the correct division ratio: He concluded that, if players A and B are a and b rounds short of winning, respectively, then the division ratio between A and B should be $b(b + 1)$: $a(a + 1)$. In our case, $a = 1$, $b = 3$, giving a division ratio of 6:1.

Pascal was at first unsure of his own solution to the problem and turned to a friend, the mathematician Gilles Personne de Roberval (1602–1675). Roberval was not of much help, and Pascal then asked for the opinion of Fermat, who was immediately intrigued by the problem. A beautiful account of the ensuing correspondence between Pascal and Fermat can be found in a recent book by Keith Devlin, *The Unfinished Game: Pascal, Fermat and the Seventeenth Century Letter That Made the World Modern* (2008). An English translation of the extant letters can be found in Smith (1929, pp. 546–565).

Fermat made use of the fact that the solution depended not on how many rounds each player had already won but *on how many each player must still win to win the prize*. This is the same observation Cardano had previously made, although he had been unable to solve the problem correctly. The solution we provided earlier is based on Fermat's idea of extending the unfinished game. Fermat also enumerated the different sample points as in our solution and reached the correct division ratio of 7:1.

Pascal seems to have been aware of Fermat's method of enumeration (Edwards 1982), at least for two players. However, when he received Fermat's method, Pascal made two important observations in his August 24, 1654, letter. First, he stated that his friend Roberval believed that there was a fault in Fermat's reasoning and that he had tried to convince Roberval that Fermat' s method was indeed correct. Roberval's argument was that, in our example, it made no sense to consider three hypothetical additional rounds, because in fact the game could end in one, two, or perhaps three rounds. The difficulty with Roberval's reasoning is that it leads us to write the sample space as in (5). Since there are three ways out of four for A to win, a naïve application of the classical definition of probability results in the wrong division ratio of 3:1 for A and B (instead of the correct 7:1). The problem here is that the sample points in Ω above are not all equally likely, so that the classical definition cannot be applied. It is thus important to consider the *maximum* number of hypothetical rounds, namely three, for us to be able to write the sample space in terms of equally likely sample points, as in Equation (5), from which the correct division ratio of 7:1 can deduced.

Pascal's second observation concerns his own belief that Fermat's method was not applicable to a game with three players. In a letter dated August 24, 1654, Pascal says (Smith 1929, p. 554),

> When there are but two players, your theory which proceeds by combinations is very just. But when there are three, I believe I have a proof that it is unjust that you should proceed in any other manner than the one I have.

Let us explain how Pascal made a slip when dealing with the Problem of Points with three players. Pascal considers the case of three players A, B, and C, who were respectively 1, 2, and 2 rounds short of winning. In this case, the maximum of further rounds before the game has to finish is $(1 + 2 + 2) - 2 = 3$.[14] With three maximum rounds, there are $3^3 = 27$ possible combinations in which the three players can win each round. Pascal correctly enumerates all the 27 ways but now makes a mistake: He counts the number of favorable combinations which lead to A winning the game as 19. As can be seen in Table 2, there are 19 combinations (denoted by check marks and Xs) for which A wins at least one round. But out of these, only 17 lead to A winning the game (the check marks) because in the remaining two (the Xs), either B or C wins the game first. Similarly, Pascal incorrectly counts the number of favorable combinations leading to B and C winning as 7 and 7, respectively, instead of 5 and 5. Pascal thus reaches an incorrect division ratio of 19:7:7.

Now Pascal again reasons incorrectly and argues that out of the 19 favorable cases for A winning the game, six of these (namely $A_1B_2B_3$, $A_1C_2C_3$, $B_1A_2B_3$, $B_1B_2A_3$, $C_1A_2C_3$, and $C_1C_2A_3$) result in either both A and B winning the game or both A and C winning the game. So he argues the net number of favorable combinations for A should be $13 + (6/2) = 16$. Likewise, he changes the number of favorable combinations for B and C, finally reaching a division ratio of $16 : 5\frac{1}{2} : 5\frac{1}{2}$. But he correctly notes

TABLE 2.
The possible combinations when A, B, and C are 1, 2, and 2 rounds short of winning the game, respectively. The check marks and Xs indicate the combinations that Pascal incorrectly chose to correspond to A winning the game. However, the Xs cannot be winning combinations for A because $B_1B_2A_3$ results in B winning and $C_1C_2A_3$ results in C winning

$A_1A_2A_3$ ✓	$B_1A_2A_3$ ✓	$C_1A_2A_3$ ✓
$A_1A_2B_3$ ✓	$B_1A_2B_3$ ✓	$C_1A_2B_3$ ✓
$A_1A_2C_3$ ✓	$B_1A_2C_3$ ✓	$C_1A_2C_3$ ✓
$A_1B_2A_3$ ✓	$B_1B_2A_3$ ×	$C_1B_2A_3$ ✓
$A_1B_2B_3$ ✓	$B_1B_2B_3$	$C_1B_2B_3$
$A_1B_2C_3$ ✓	$B_1B_2C_3$	$C_1B_2C_3$
$A_1C_2A_3$ ✓	$B_1C_2A_3$ ✓	$C_1C_2A_3$ ×
$A_1C_2B_3$ ✓	$B_1C_2B_3$	$C_1C_2B_3$
$A_1C_2C_3$ ✓	$B_1C_2C_3$	$C_1C_2C_3$

that the answer cannot be right, for his own recursive method gives the correct ratio of 17:5:5. Thus, Pascal at first wrongly believed Fermat's method of enumeration was not generalizable to more than two players. Fermat was quick to point out the error in Pascal's reasoning. In his letter dated September 25, 1654, Fermat explains (Smith 1929, p. 562),

> In taking the example of the three gamblers of whom the first lacks one point, and each of the others lack two, which is the case in which you oppose, I find here only 17 combinations for the first and 5 for each of the others; for when you say that the combination *acc* is good for the first, recollect that everything that is done after one of the players has won is worth nothing. But this combination having made the first win on the first die, what does it matter that the third gains two afterwards, since even when he gains thirty all this is superfluous? The consequence, as you have well called it "this fiction," of extending the game to a certain number of plays serves only to make the rule easy and (according to my opinion) to make all the chances equal; or better, more intelligibly to reduce all the fractions to the same denomination.

We next move to the renowned German mathematician and philosopher Gottfried Wilhelm Leibniz (1646–1716), who is usually remembered as the coinventor of differential calculus, with archrival Isaac Newton. However, he was also interested in probability and famously made a similar mistake of incorrectly enumerating sample points. When confronted with the question "With two dice, is a throw of twelve as likely as a throw of eleven?" Leibniz states in the *Opera Omnia* (Leibniz 1768, p. 217),

> . . . for example, with two dice, it is equally likely to throw twelve points, than to throw eleven; because one or the other can be done in only one manner.

Thus, Leibniz believed the two throws to be equally likely, arguing that in each case the throw could be obtained in a single way. Although it is true that a throw of 11 can be realized only with a five and a six, there are two ways in which it could happen: the first die could be a five and the second a six, or vice versa. On the other hand, a throw of 12 can be realized in only one way: a six on each die. Thus the first probability is twice the second. Commenting on Leibniz's error, Todhunter states (Todhunter 1865, p. 48),

> Leibniz however furnishes an example of the liability to error which seems peculiarly characteristic of our subject.

Nonetheless, this should not in any way undermine some of the contributions Leibniz made to probability theory. For one thing, he was one of the very first to give an explicit definition of classical probability, except phrased in terms of an expectation (Leibniz 1969, p. 161),

> If a situation can lead to different advantageous results ruling out each other, the estimation of the expectation will be the sum of the possible advantages for the set of all these results, divided into the total number of results.

In spite of being conversant with the classical definition, Leibniz was interested in establishing a logical theory for different degrees of certainty. He may rightly be regarded as a precursor to later developments in the logical foundations of probability by Keynes, Jeffreys, Carnap, and others. Since Jacob Bernoulli had similar interests, Leibniz started a communication with him in 1703. He undoubtedly had some influence in Bernoulli's *Ars Conjectandi* (Bernoulli 1713). When Bernoulli communicated to Leibniz about his law of large numbers, the latter reacted critically. As Schneider explains (2005, p. 90),

> Leibniz's main criticisms were that the probability of contingent events, which he identified with dependence on infinitely many conditions, could not be determined by a finite number of observations and that the appearance of new circumstances could change the probability of an event. Bernoulli agreed that only a finite number of trials can be undertaken; but he differed from Leibniz in being convinced by the urn model that a reasonably great number of trials yielded estimates of the sought-after probabilities that were sufficient for all practical purposes.

Thus, in spite of Leibniz's criticism, Bernoulli was convinced of the authenticity of his theorem. This situation is fortunate because Bernoulli's law was nothing less than a watershed moment in the history of probability.

A few years after Leibniz's death, Jean le Rond d'Alembert (1717–1783), who was one of the foremost intellectuals of his times, infamously considered the following problem: "In two tosses of a fair coin, what is the probability that heads will appear at least once?" For this problem, d'Alembert denied that 3/4 could be the correct answer. He reasoned as follows: once a head occurs, there is no need for a second throw; the possible outcomes are thus *H*, *T H*, *T T*, and the required probability is 2/3. Of course, d'Alembert's reasoning is wrong because he failed to realize that each of *H*, *T H*, *T T* is *not* equally likely. The erroneous answer was even included in his article *Croix ou Pile*[15] of the *Encyclopédie* (d'Alembert 1754, Vol. IV, pp. 512–513). Bertrand (1889, pp. ix–x) did not mince his words about d'Alembert's various faux pas in the games of chance:

> When it comes to the calculus of probability, D'Alembert's astute mind slips completely.

Similarly, in his *History of Statistics,* Karl Pearson writes (Pearson 1978, p. 535),

> What then did D'Alembert contribute to our subject? I think the answer to that question must be that he contributed absolutely *nothing*.

In spite of Bertrand's and Pearson's somewhat harsh words, it would be misleading for us to think that d'Alembert, a man of immense mathematical prowess, was so naïve that he would have no strong basis for his probabilistic reasoning. In the *Croix ou Pile* article, a sample space of $\{H\,H, H\,T, T\,H, T\,T\}$ made no sense to d'Alembert because it did not correspond to reality. In real life, no person would ever observe $H\,H$, because once an initial H was observed the game would end. By proposing an alternative model for the calculus of probabilities, namely that of equiprobability on *observable* events, d'Alembert was effectively asking why his model could not be right, given the absence of an existing theoretical framework for the calculus of probabilities. D'Alembert's skepticism was partly responsible for later mathematicians seeking a solid theoretical foundation for probability, culminating in its axiomatization by Kolmogorov (1933).

4. Confusion Regarding the Use of Statistical Independence

D'Alembert also famously considered the following problem: "When a fair coin is tossed, given that heads have occurred three times in a row, what is the probability that the next toss is a tail?" When presented with the problem, d'Alembert insisted that the probability of a tail must "obviously" be greater than 1/2,[16] thus rejecting the concept of independence between the tosses. The claim was made in d'Alembert's *Opuscules Mathématiques* (d'Alembert 1761, pp. 13–14). In his own words,

> Let's look at other examples which I promised in the previous Article, which show the lack of exactitude in the ordinary calculus of probabilities.
>
> In this calculus, by combining all possible events, we make two assumptions which can, it seems to me, be contested. The first of these assumptions is that, if an event has occurred several times successively, for example, if in the game of heads and tails, heads has occurred three times in a row, it is equally likely that head or tail will occur on the fourth time? However I ask if this assumption is really true, and if the number of times that heads has already successively occurred by the hypothesis, does not make it

more likely the occurrence of tails on the fourth time? Because after all it is not possible, it is even physically impossible that tails never occurs. Therefore the more heads occurs successively, the more it is likely tail will occur the next time. If this is the case, as it seems to me one will not disagree, the rule of combination of possible events is thus still deficient in this respect.

D'Alembert states that it is *physically* impossible for tails never to occur in a long series of tosses of a coin, and thus used his concepts of physical and metaphysical probabilities[17] to support his erroneous argument.

D'Alembert's remarks need some clarification because the misconceptions are still widely believed. Consider the following two sequences when a fair coin is tossed four times:

sequence 1 : $H\ H\ H\ H$
sequence 2 : $H\ H\ H\ T$

Many would believe that the first sequence is less likely than the second one. After all, it seems highly improbable to obtain four heads in a row. However, it is equally unlikely to obtain the second sequence *in that specific order.* Although it is less likely to obtain four heads than to obtain a total of three heads and one tail,[18] $H\ H\ H\ T$ is as likely as any other of the same length, even if it contains all heads or all tails.

A more subtle "mistake" concerning the issue of independence was made by Laplace. Pierre-Simon Laplace (1749–1827) was a real giant in mathematics. His works on inverse probability were fundamental in bringing the Bayesian paradigm to the forefront of the calculus of probability and of statistical inference. Hogben says (1957, p. 133),

> The *fons et irigo* of inverse probability is Laplace. For good or ill, the ideas commonly identified with the name of Bayes are largely his.

Indeed, the form of Bayes' theorem as it usually appears in textbooks, namely

$$\Pr\{A_j \mid B\} = \frac{\Pr\{B \mid A_j\} \Pr\{A_j\}}{\sum_{i=1}^{n} \Pr\{B \mid A_i\} \Pr\{A_i\}} \tag{6}$$

is due to Laplace. In Equation (6), $A_1, A_2, \ldots, A_n$ is a sequence of mutually exclusive and exhaustive events, $\Pr\{A_j\}$is the prior probability of A_j, and $\Pr\{A_j|B\}$ is the posterior probability of A_j given B. The continuous version of Equation (6) can be written as

$$f(\theta|\mathbf{x}) = \frac{f(\mathbf{x}|\theta)f(\theta)}{\int_{-\infty}^{\infty} f(\mathbf{x}|\theta)f(\theta)d\theta}$$

where $f(\theta)$ is the prior density of θ, $f(\mathbf{x}|\theta)$ is the likelihood of the data $\mathbf{x}$, and $f(\theta|\mathbf{x})$ is the posterior density of θ.

Before commenting on a specific example of Laplace's work on inverse probability, let us recall that it is with him that the classical definition of probability is usually associated, for he was the first to have given it in its clearest terms. Indeed, Laplace's classical definition of probability is the one that is still used today. In his very first paper on probability, *Mémoire sur les suites récurro-recurrentes et sur leurs usages dans la théorie des hasards* (Laplace 1774b), Laplace writes,

> . . . if each case is equally probable, the probability of the event is equal to the number of favorable cases divided by the number of all possible cases.

This definition was repeated both in Laplace's *Théorie Analytique* and *Essai Philosophique* (1814).

The rule in Equation (6) was first enunciated by Laplace in his 1774 *Mémoire de la Probabilité des Causes par les Evènements* (Laplace 1774a). This is how Laplace phrases it:

> If an event can be produced by a number *n* of different causes, the probabilities of the existence of these causes, calculated from the event, are to each other as the probabilities of the event, calculated from the causes; and the probability of each cause is equal to the probability of the event, calculated from that cause, divided by the sum of all the probabilities of the event, calculated from each of the causes.

It is very likely that Laplace was unaware of Bayes' previous work on inverse probability (Bayes 1764) when he enunciated the rule in 1774. However, the 1778 volume of the *Histoire de l'Académie Royale des Sciences*, which appeared in 1781, contains an interesting summary by the Marquis de Condorcet[19] (1743–1794) of Laplace's article *Sur les Probabilités*, which also appeared in that volume (Laplace 1781). Although Laplace's article itself makes mention of neither Bayes nor Price,[20] Condorcet's summary explicitly acknowledges the two Englishmen[21] (Laplace 1781, p. 43):

> These questions [on inverse probability] about which it seems that Messrs. Bernoulli and Moivre had thought, have been since then examined by Messrs. Bayes and Price; but they have limited themselves to exposing the principles that can be used to solve them. M. de Laplace has expanded on them . . .

Coming back to the 1774 paper, after having enunciated his principle on inverse probability, Laplace is famous for discussing the following problem: "A box contains a large number of black and white balls. We sample n balls with replacement, of which b turn out to be black and $n - b$ turn out to be white. What is the conditional probability that the next ball drawn will be black?" Laplace's solution to this problem essentially boils down to the following, in modern notation. Let X_n be the number of black balls out of the sample of size n, and let the probability that a ball is black be p. Also, let B^* be the event that the next ball is black. From Bayes' theorem, we have

$$\begin{aligned} f(p \mid X_n = b) &= \frac{\Pr\{X_n = b \mid p\} f(p)}{\Pr\{X_n = b\}} \\ &= \frac{\Pr\{X_n = b \mid p\} f(p)}{\int_0^1 \Pr\{X_n = b \mid p\} f(p)\, dp} \end{aligned}$$

Then the required probability is

$$\begin{aligned} \Pr\{B^* \mid X_n = b\} &= \int_0^1 \Pr\{B^* \mid p, X_n = b\} f(p \mid X_n = b)\, dp \\ &= \frac{\int_0^1 p \cdot \Pr\{X_n = b \mid p\} f(p)\, dp}{\int_0^1 \Pr\{X_n = b \mid p\} f(p)\, dp} \end{aligned}$$

In the above, it is assumed that $\Pr\{B^* \mid p, X_n = b\} = p$, that is, each ball is drawn independently of the other. Laplace also assumes that p is uniform in [0,1], so that

$$\Pr\{B^* \mid X_n = b\} = \frac{\int_0^1 p^{b+1}(1-p)^{n-b}\, dp}{\int_0^1 p^b (1-p)^{n-b}\, dp} = \frac{b+1}{n+2}$$

In particular, if all of the n balls turn out to be black, then the probability that the next ball is also black is $(n + 1)/(n + 2)$. The above problem has been much discussed in the literature and is known as Laplace's rule of succession.[22] Using the rule of succession, Laplace considered the following question: "Given that the sun has risen every day for the past 5,000 years, what is the probability that it will rise tomorrow?" Substituting $n = 5{,}000 \times 365.2426 = 1{,}826{,}213$ in the above formula, Laplace obtained the probability 1,826,214/1,826,215 (0.9999994). Thus, in his *Essai Philosophique sur les Probabilités*[23] (1814) English edition, p. 19, Laplace says,

> Thus we find that an event having occurred successively any number of times, the probability that it will happen again the next time is equal to this number increased by unity divided by the same number, increased by two units. Placing the most ancient epoch of history at five thousand years ago, or at 1,826,213 days, and the sun having risen constantly in the interval at each revolution of 24 hours, it is a bet of 1,826,214 to one that it will rise again tomorrow.

Laplace's calculation was meant to be an answer to *Hume's problem of induction*. Fifteen years before the publication of Bayes' *Essay*, the eminent Scottish philosopher David Hume (1711–1776) wrote his groundbreaking book *An Enquiry Concerning Human Understanding* (Hume 1748). In this work, Hume formulated his famous *problem of induction*, which we now explain. Suppose out of a large number n of occurrences of an event A, an event B occurs m times. Based on these observations, an inductive inference would lead us to believe that approximately m/n of all events of type A is also of type B, that is, the probability of B given A is approximately m/n. Hume's problem of induction states that such an inference has no rational justification but arises only as a consequence of custom and habit. Earlier in his book, Hume gave the famous "rise-of-the sun" example, which was meant to illustrate the shaky ground on which "matters of fact" or inductive reasoning rested (Hume 1748):

> Matters of fact, which are the second objects of human reason, are not ascertained in the same manner; nor is our evidence of their truth, however great, of a like nature with the foregoing. The contrary of every matter of fact is still possible; because it can

> never imply a contradiction, and is conceived by the mind with the same facility and distinctness, as if ever so conformable to reality. That the sun will not rise to-morrow is no less intelligible a proposition, and implies no more contradiction, than the affirmation, that it will rise. We should in vain, therefore, attempt to demonstrate its falsehood. Were it demonstratively false, it would imply a contradiction, and could never be distinctly conceived by the mind.

Laplace thus thought that his calculations provided a possible solution to Hume's problem of induction. However, Laplace, who so often has been called France's Newton, was harshly criticized for his calculations. Zabell says (2005, p. 47),

> Laplace has perhaps received more ridicule for this statement than for any other.

Somehow, Laplace must have felt that there was something amiss with his calculations. For his very next sentence reads,

> But this number is incomparably greater for him who, recognizing in the totality of phenomena the principal regulator of days and seasons, sees that nothing at the present moment can arrest the course of it.

Laplace here seems to warn the reader that his method is correct when based only on the information from the sample, but his statement is too timid. To understand the criticism leveled against Laplace's calculation, consider the following example given by the Austro-British philosopher Karl Popper (1902–1994) (Popper 1957; Gillies 2000, p. 73): Suppose that the sun rises for 1,826,213 days (5,000 years), but then suddenly the Earth stops rotating on day 1,826,214. Then, for parts of the globe (say Part A), the sun does not rise on that day, whereas for other parts (say Part B), the sun will appear fixed in the sky. What then is the probability that the sun will rise again in Part A of the globe? Applying the generalized form of the rule of succession with $n = 1{,}826{,}214$ and $B = 1{,}826{,}213$ gives a probability of 0.9999989, which is almost as high as the original probability of 0.9999994! The answer is preposterous since it does not give enough weight to the recent failure.

The rule of succession is perfectly valid as long as the assumptions it makes are all tenable. Applying the rule of succession to the rising of the sun, however, should be viewed with skepticism for several reasons (see, e.g., Schay 2007, p. 65). A major criticism lies in the assumption of independence. Moreover, it is also dubious that the rising of the sun on a given day can be considered a random event at all. Finally, the solution relies on the principle of indifference: The probability of the sun rising is equally likely to take any of the values in [0,1] because there is no reason to favor any particular value for the probability. To many, this is not a reasonable assumption.

5. Conclusion

We have outlined some of the more well-known errors that were made during the early development of the theory of probability. The solution to the problems we considered would seem quite elementary nowadays. It must be borne in mind, however, that in the times of those considered here and even afterwards, notions about probability, sample spaces, and sample points were quite abstruse. It took a while before the proper notion of a mathematical model was developed, and a proper axiomatic model of probability was developed only as late as 1933 by Kolmogorov (1933). Perhaps, then, the personalities and their errors discussed in this article should not be judged too harshly.

Notes

1. *The Book on Games of Chance.* An English translation of the book and a thorough analysis of Cardano's connections with games of chance can be found in Ore's *Cardano: The Gambling Scholar* (Ore 1953). More bibliographic details can be found in Gliozzi (1980, pp. 64–67) and Scardovi (2004, pp. 754–758).

2. The correct answer is four and can be obtained by solving for the smallest N integer such that $1 - (5/6)^n = 1/2$.

3. Cardano frequently uses the term "equality" in the *Liber* to denote half of the total number of sample points in the sample space. See Ore (1953, p. 149).

4. Real name Antoine Gombaud. Leibniz describes the Chevalier de Méré as "a man of penetrating mind who was both a player and a philosopher" (Leibniz 1896, p. 539). Pascal biographer Tulloch also notes (1878, p. 66): "Among the men whom Pascal evidently met at the hotel of the Duc de Roannez [Pascal's younger friend], and with whom he formed something of a friendship, was the well-known Chevalier de Méré, whom we know best as a tutor of Madame de Maintenon, and whose graceful but flippant letters still survive as a picture of the time. He was a gambler and libertine, yet with some tincture of science and professed

interest in its progress." Pascal himself was less flattering. In a letter to Fermat, Pascal says (Smith 1929, p. 552): ". . . he [de Méré] has ability but he is not a geometer (which is, as you know, a great defect) and he does not even comprehend that a mathematical line is infinitely divisible and he is firmly convinced that it is composed of a finite number of points. I have never been able to get him out of it. If you could do so, it would make him perfect." The book by Chamaillard (1921) is completely devoted to the Chevalier de Méré.

5. Ore (1960) believes that the difference in the probabilities for 24 and 25 throws is so small that it is unlikely that de Méré could have detected this difference through observations.

6. Of the several books that have been written on Pascal, the biographies by Groothuis (2003) and Hammond (2003) are good introductions to his life and works.

7. Carcavi had been an old friend of Pascal's father and was very close to Pascal.

8. Fermat is today mostly remembered for the so-called "Fermat's Last Theorem," which he conjectured in 1637 and which was not proved until 1995 by Andrew Wiles (1995). The theorem states that no three positive integers a, b, c can satisfy the equation $a^n + b^n = c^n$ for any integer n greater than 2. A good introduction to Fermat's Last Theorem can be found in Aczel (1996). The book by Mahoney (1994) is an excellent biography of Fermat, whose probability work appears on pp. 402–410 of the book.

9. For example, if we apply de Moivre's Gambling Rule to the one-die problem, we obtain $x = 0.693/(1/6) = 4.158$ so that $C = 5$. This answer cannot be correct because we showed in the solution that we need only four tosses.

10. The Problem of Points is also discussed by Todhunter (1865, Chap. II), Hald (2003, pp. 56–63), Petkovic (2009, pp. 212–214), Paolella (2006, pp. 97– 99), Montucla (1802, pp. 383–390), Marques de Sá (2007, pp. 61–62), Kaplan and Kaplan (2006, pp. 25–30), and Isaac (1995, p. 55).

11. For a full discussion of the Problem of Points before Pascal, see Coumet (1965).

12. *Everything about Arithmetic, Geometry, and Proportion.*

13. The correct division ratio for A and B here is approximately 53:47.

14. The general formula is: Maximum number of remaining rounds = (sum of the number of rounds each player is short of winning) − (number of players − 1).

15. *Heads or Tails.*

16. The correct answer is, of course, 1/2.

17. According to d'Alembert, an event is metaphysically possible if its probability is greater than zero and is physically possible if it is not so rare that its probability is very close to zero.

18. Remember that the specific sequence *H H H T* is one of four possible ways of obtaining a total of three heads and one tail.

19. Condorcet was assistant secretary in the Académie des Sciences and was in charge of editing Laplace's papers for the transactions of the academy.

20. Upon Bayes' death, his friend Richard Price (1723–1791) decided to publish some of his papers with the Royal Society. Bayes' *Essay* (1764) was augmented by an introduction and an appendix written by Price.

21. Laplace's acknowledgment of Bayes appears in his *Essai Philosophique* (Laplace 1814) English edition, p. 189.

22. Laplace's rule of succession is also discussed by Pitman (1993, p. 421), Sarkar and Pfeifer (2006, p. 47), Pearson (1900, pp. 140–150), Zabell (2005,Chap. 2), Jackman (2009, p. 57), Keynes (1921, p. 376), Chatterjee (2003, pp. 216–218), Good (1983, p. 67), Gelman et al. (2003, p. 36), Blom et al. (1994, p. 58), Isaac (1995, p. 36), and Chung and AitSahlia (2003, p. 129).

23. *Philosophical Essay on Probabilities.*

References

Aczel, A. D. (1996), *Fermat's Last Theorem: Unlocking the Secret of an Ancient Mathematical Problem*, Delta Trade Paperbacks.

Bayes, T. (1764), "An Essay Towards Solving a Problem in the Doctrine of Chances," *Philosophical Transactions of the Royal Society of London*, 53, 370–418. Reprinted in *Studies in the History of Statistics and Probability*, Vol. 1, eds. E. S. Pearson and M. G. Kendall, London: Charles Griffin, 1970, pp. 134–153.

Bernoulli, J. (1713), *Ars Conjectandi*, Basel.

Bertrand, J. (1889), *Calcul des Probabilités*, Gauthier-Villars et fils, Paris.

Blom, G., Holst, L., and Sandell, D. (1994), *Problems and Snapshots from the World of Probability*, Berlin: Springer-Verlag.

Burton, D. M. (2006), *The History of Mathematics: An Introduction* (6th ed.), New York: McGraw-Hill.

Cardano, G. (1539), *Practica Arithmetice, & Mensurandi Singularis. In qua que preter alias cōtinentur, versa pagina demonstrabit*, Io. Antonins Castellioneus medidani imprimebat, impensis Bernardini calusci., Milan (Appears as Practica Arithmeticae Generalis Omnium Copiosissima & Utilissima, in the 1663 ed.).

——— (1564), *Liber de Ludo Aleae*, first printed in Opera Omnia, Vol. 1, 1663 Edition (pp. 262–276).

——— (1935), *Ma Vie*, Paris (translated by Jean Dayre).

Chamaillard, E. (1921), *Le Chevalier de Méré*, G. Clouzot, Niort.

Chatterjee, S. K. (2003), *Statistical Thought: A Perspective and History*, Oxford: Oxford University Press.

Chung, K. L., and AitSahlia, F. (2003), *Elementary Probability Theory* (4th ed.), Berlin: Springer-Verlag.

Coumet, E. (1965), "Le Problème des Partis avant Pascal," *Archives Internationales d'Histoire des Sciences*, 18, 245–272.

d'Alembert, J. L. R. (1754), "Croix ou Pile," in *Encyclopédie ou Dictionnaire Raisonné des Sciences, des Arts et des Métiers* (Vol. 4), eds. D. Diderot and J. L. R. d'Alembert, Stuttgart.

——— (1761), *Opuscules Mathématiques* (Vol. 2), Paris: David.

de Moivre, A. (1718), *The Doctrine of Chances, or a Method of Calculating the Probability of Events in Play* (1st Ed.), London: Millar.

Devlin, K. (2008), *The Unfinished Game: Pascal, Fermat, and the Seventeenth-Century Letter That Made the World Modern*, New York: Basic Books.

Edwards, A. W. F. (1982), "Pascal and the Problem of Points," *International Statistical Review*, 50, 259–266.

Gelman, A., Carlin, J. B., Stern, H. S., and Rubin, D. (2003), *Bayesian Data Analysis* (2nd Ed.), Boca Raton, Fl.: Chapman & Hall/CRC.

Gillies, D. A. (2000), *Philosophical Theories of Probability*, London: Routledge.

Gliozzi, M. (1980), "Cardano, Girolamo," in *Dictionary of Scientific Biography* (Vol. 3), ed. C. C. Gillispie, New York: Charles Scribner's Sons.

Good, I. J. (1983), *Good Thinking: The Foundations of Probability and Its Applications*, Minneapolis: University of Minnesota Press.

Groothuis, D. (2003), *On Pascal*, Belmont, Calif.: Thomson Wadsworth.

Hald, A. (2003), *A History of Probability and Statistics and Their Applications Before 1750*, New York: Wiley.

Hammond, N. (ed.) (2003), *The Cambridge Companion to Pascal*, Cambridge: Cambridge University Press.

Hogben, L. (1957), *Statistical Theory: The Relationship of Probability, Credibility and Error*, New York: W.W. Norton & Co.

Hume, D. (1748), *An Enquiry Concerning Human Understanding*, ed. P. Millican. London (2007 edition edited by P. Millican, Oxford University Press, Oxford).

Isaac, R. (1995), *The Pleasures of Probability*, New York: Springer-Verlag.

Jackman, S. (2009), *Bayesian Modelling in the Social Sciences*, New York: Wiley.

Kaplan, M., and Kaplan, E. (2006), *Chances Are: Adventures in Probability*, Baltimore: Penguin Books.

Keynes, J. M. (1921), *A Treatise on Probability*, London: Macmillan & Co.

Kolmogorov, A. (1933), *Grundbegriffe der Wahrscheinlichkeitsrechnung*, Berlin: Springer.

Laplace, P.-S. (1774a), "Mémoire de la Probabilité des Causes par les Evènements," *Memoire de l'Académie Royale des Sciences de Paris (savants étrangers)*, Tome VI: 621–656.

——— (1774b), "Mémoire sur les Suites Récurro-récurrentes et sur leurs usages dans la théorie des hasards," *Mémoire de l'Académie Royale des Sciences de Paris (savants étrangers)* Tome VI: 353–371.

——— (1781), "Sur les probabilités," *Histoire de l'Académie Royale des Sciences*, année 1778, Tome VI: 227–323.

——— (1814), *Essai Philosophique sur les Probabilités*, Paris: Courcier: Paris (6th ed. 1840, translated 1902 as *A Philosophical Essay on Probabilities*, translated by F. W. Truscott & F. L. Emory. Reprinted, Dover, New York, 1951).

Leibniz, G. W. (1768), *Opera Omnia*, Geneva.

——— (1896), *New Essays Concerning Human Understanding*, New York: Macmillan (original work written in 1704 and published in 1765).

——— (1969), *Théodicée*, Paris: Garnier-Flammarion (original work published in 1710).

Mahoney, M. S. (1994), *The Mathematical Career of Pierre de Fermat, 1601–1665* (2nd Ed.), Princeton, N.J.: Princeton University Press.

Marques de Sá, J. P. (2007), *Chance: The Life of Games & the Game of Life*, Berlin: Springer-Verlag.

Montucla, J. F. (1802), *Histoire des Mathématiques* (Tome III), Paris: Henri Agasse.

Ore, O. (1953), *Cardano: The Gambling Scholar*, Princeton, N.J.: Princeton University Press.

——— (1960), "Pascal and the Invention of Probability Theory," *American Mathematical Monthly*, 67, 409–419.

Pacioli, L. (1494), *Summa de Arithmetica, Geometrica, Proportioni, et Proportionalita*, Venice.

Paolella, M. S. (2006), *Fundamental Probability: A Computational Approach*, New York: Wiley.

Pearson, E. S. (ed.) (1978), *The History of Statistics in the 17th and 18th Centuries, Against the Changing Background of Intellectual, Scientific and Religious Thought. Lectures by Karl Pearson Given at University College London During Academic Sessions 1921–1933*, London: Griffin.

Pearson, K. (1900), *The Grammar of Science*, (2nd Ed.), London: Adam and Charles Black.

Petkovic, M. S. (2009), *Famous Puzzles of Great Mathematicians*, American Mathematical Society.

Pitman, J. (1993), *Probability*, New York: Springer.

Popper, K. R. (1957), "Probability Magic or Knowledge out of Ignorance," *Dialectica*, 11, 354–374.

Sarkar, S., and Pfeifer, J. (2006), *The Philosophy of Science: An Encyclopedia*, London: Routledge.

Scardovi, I. (2004), "Cardano, Gerolamo," in *Encyclopedia of Statistical Sciences* (2nd ed.), eds. S. Kotz, C. B. Read, N. Balakrishnan, and B. Vidakovic, New York: Wiley.

Schay, G. (2007), *Introduction to Probability with Statistical Applications*, Boston: Birkhauser.

Schneider, I. (2005), "Jakob Bernoulli, *Ars Conjectandi* (1713)," in *Landmark Writings in Western Mathematics 1640–1940*, ed. I. Grattan-Guinness, Amsterdam: Elsevier.

Smith, D. E. (1929), *A Source Book in Mathematics*, New York: McGraw-Hill.

Todhunter, I. (1865), *A History of the Mathematical Theory of Probability From the Time of Pascal to That of Laplace*, London: Macmillan (Reprinted by Chelsea, New York, 1949, 1965).

Tulloch, P. (1878), *Pascal*, London: William Blackwood and Sons.

Weaver, W. (1982), *Lady Luck: The Theory of Probability*, New York: Dover (originally published by Anchor Books, Doubleday & Company, Inc., Garden City, N.Y., 1963).

Wiles, A. (1995), "Modular Elliptic Curves and Fermat's Last Theorem," *The Annals of Mathematics*, 141, 443–551.

Williams, L. (2005), "Cardano and the Gambler's Habitus," *Studies in History and Philosophy of Science*, 36, 23–41.

Zabell, S. L. (2005), *Symmetry and its Discontents*, Cambridge: Cambridge University Press.

The End of Probability

Elie Ayache

Contingency

Probability theory and the metaphysical category of possibility are based on the notion of "states of the world" (or possible worlds). In the market, the only states of the world are prices. Contingency is a very general category that is independent of the later division of the world into identifiable states or the recognition of the different possible worlds that the world might be. Metaphysical thought *later* works contingency into the notion of separable possible states. However, pure and absolute (and initial) contingency only minimally says that the world or that the things in it could have been different.

A contingent claim is generally a claim that pays out something but that could pay something different (following the minimal definition of the contingent world as a world that is so but that could be different). Because the only thing that can make a difference in the market are prices, and because different prices are the only thing that can suggest that the actual world could have been different, we usually define the difference of the contingent claim in terms of underlying prices: if S is greater than K, pay \$1; else, pay 0 (in the typical case of the contingent claim known as the digital option, where S is the price of the underlying, typically the price of the underlying share).

In the real world, contingency is not reducible to underlying states because truly unpredictable events are typically those that escape the previously known range of possibilities or possible states. The current saying is that true events create the possibilities that will have led to them. It is out of the question to apply probability to those events, or even to say that their probability was zero because probability is only defined relative to a stable and well-defined universe of possibilities.

It is our metaphysical tendency to reduce contingency to identifiable possible states that makes our models and view of the world vulnerable to their outside, or to contingencies that were not previously identified. Consequently we ask: Why not drop the division into states (i.e., possibility) altogether and deal with contingency pure and simple (i.e., without states)?

The Market

Possible states and probability theory are pervasive in the market because we feel confident that the underlying states there are prices anyway, so the range of possibilities seems to be identified and totalized once and for all. What radical change could contingency bring over and above prices? No matter how many unpredictable events hit the market, the "world" of the underlying share would always consist of the array of prices of that share, wouldn't it? Because the market is composed of numbers (prices), we feel confident applying probability to it.

However, precisely because of the precision of numbers, the market is precisely the domain where probability and states of the world demonstrably fail in front of the category of price, which, in my view, is supposed to replace probability and states of the world.

Indeed, if states of the world in the market are prices, then the prices of the contingent claims should also be states of the world, different from those of the underlying. This possibility is what probability theory, and its culmination—which is derivative valuation theory—cannot allow. According to derivative valuation theory, the derivative value is a deterministic function of the underlying price. In the framework of Black–Scholes–Merton (BSM), option value is a deterministic function of the underlying and options are redundant. However, everybody knows that options trade independently of their underlying in a market of their own, thus adding new states of the world (in other words, trading options turns the BSM volatility stochastic). The purpose of writing any contingent claim is to trade it and to add a new price to the market. Yet the framework of fixed states of the world and of probability automatically leads to the dynamic replication of the contingent claim, and hence to its redundancy.

The only way out is to conceive of contingent claims and their market prices independently of the whole framework of possibility and underlying states of the world. Surely, the price of a contingent claim only depends on its underlying at maturity, and this is the only reason why we call it a "derivative." This is the only reason why derivative valuation theory, which is just the mathematical exploration of the terminal dependence of the contingent claim on its underlying, likewise calls it a "derivative." However, the market of the contingent claim is what happens *before* its maturity. (Derivative valuation theory ignores the market of contingent claims.) Before it expires, the price of the contingent claim may depend, not only on its underlying, but also on such things as the volatility of the underlying or on the volatility of its volatility. In a word, it may depend on the whole market.

We express this instant nesting (or "complication") in the language of prices by saying that vanilla options never trade alone, as their valuation theory prescribes. Barrier options, variance swaps, options on variance (options on volatility index), cliquets, etc.—all these structures trade as well at prices not necessarily predicted by the model that we had initially, candidly, thought was all we needed to value the vanillas, i.e., BSM. As a consequence, no one can claim to rightly price vanillas unless one's model is calibrated to such things as the market prices of barrier options, variance swaps, options on variance, etc. The rule is constant recalibration of the model to the novel prices of novel structures. This is the rule of the market. The market is a constant black swan: It constantly breaks any previously defined range of possibilities.

The Tree of Possibilities

We should forget about probability and stochastic processes altogether. Every day brings a new market, to which we recalibrate. Probability and stochastic process impose on us the notion of a transition. We move from one day to the next by transitioning between the given states of the world, with some probability. In reality, however, the whole idea of a transition should be severed. There is no such thing as a tree of possibilities, and consequently no transition. The market attaches prices to the contingent claims immanently, without a supervisor, without the transcendent view of states of the worlds which are supposed to distribute probability once and for all.

If the contingent claim is thought of as a derivative, i.e., as a function of its underlying, this thinking imposes on us that we chart the corresponding underlying states. Not only do we have to identify them, but by the same token, we also eliminate all the others: We reduce the contingency of the world to only those states. By contrast, if the contingent claim is only conceived as the written formula that it is (pay $1 if S is greater than K, 0 otherwise), its only underlying will be the sheet of paper on which it is written. This sheet is not divided into states (those above K and those below) and doesn't presuppose the notion of states. It doesn't impose on the world that the only states that it might experience the next day are states of the underlying S. For all we know, nuclear war might break out the next day, or the world as a whole might disappear. The sheet of paper is indifferent to all of this. It will still exist the next day, and it will still admit of a price (provided that the market still exists). By contrast, the probability of a state of the world that was identified the day before may no longer be defined the next day because of the major shift of the whole range of possibilities.

Recognizing absolute contingency is only recognizing that contingency should be considered absolutely and not derivatively on identifiable possibilities. Time passes every day, and the world is still contingent the next day, without there being an array of states of the world that mediate the transition from one day to the next. Contingent claims admit of prices one day, and they admit of prices the next day, without there being any common measure between what we identify as "states of the world" on one day and "states of the world" on the next day.

Valuation theory shackles us to the tree of possibilities, with the necessity that the states of the world of the next day be commensurate with those of the previous day because the theoretical value of the contingent claim is computed as a backward induction in the tree. We need to project all future possible states (as if the future was perfectly imaginable and wasn't absolutely incommensurate with the present) in order to compute the present value of the contingent claim as the discounted mathematical expectation of its value in the future states.

In reality, however, the market prices the contingent claims going forward. Each day brings a new price and a new market. Valuation theory seems to know of no other way but backwards. Instead of forcing the market into valuation theory and thinking of the market price as the theoretical result of some general equilibrium problem-solving algorithm, no less based on fixed and identifiable states of the world, why don't we just embrace the market as a radical alternative to valuation theory?

We just have to admit that the market finds the price of the contingent claim, immediately and immanently, by definition of the market.

Writing and Exchange

Contingency is absolute; it doesn't need the states of possibilities, or the tree of possibilities, or the transitions (all these constructs are too fragile and too unstable). Every day the world is the way it is, and every day it could have been different. To repeat, there is no visible or identifiable transition. As a matter of fact, we go to sleep in the interval and nobody knows what happens when we sleep! Contingency is indifferent to the passage of time. There should be no difference between a thing that exists already or is even past yet could have been different, and a future thing that doesn't exist, yet we know is contingent. Contingency comes before existence. By contrast, possibility is sensitive to the passage of time. A past thing is no longer possible, because it becomes actual.

In the market, there are only contingent claims. It is only incidentally that the contingency of a certain contingent claim may be defined relative to the prices of another (which is then called the underlying). As I have said, this result occurs because prices are chiefly what distinguish between different worlds. However, nothing stops us from defining contingent claims whose underlying may be such things as the weather, or earthquakes, or corporate defaults.

The market is the medium of contingent claims, and it translates them into prices. As I have said, contingent claims are recognizable by their written formula. It is because they are written that they can dispense with possibility and its underlying states of the world. Even though the written formula may have a hundred provisions, it will remain ONE written formula, which will be interpreted the next day, and the day after the next, and it will admit of a price every day. At no point do we need to decompose it into states.

Now the other side of something being written is that this something will then be exchanged. It is not a coincidence if writing enables us to collect the contingent claim in one undivided formula and if, on the other hand, it lends itself to the exchange. It is the same matter (the material sheet) on which the formula is written which is subsequently exchanged. This notion might even act as the definition of writing: something that collects the difference of the contingent claim on one side and for this reason admits of a price (i.e., is exchanged) on the other side.

Probability theory thinks only of the stochastic process of the underlying (i.e., underlying states and the corresponding probabilistic transitions), and the value of the derivative is then tracked by stochastic control (i.e., dynamic replication or dynamic programming). However, all of this valuation is not real. What really exist are contingent claims, their market, and their prices. Nobody ever observes a stochastic process. We are not even sure what probability is! The only pricing technology that is worth having is a tool that can be calibrated and recalibrated to the market prices, without any presumption of states. The only reason why probability and stochastic control are episodically needed is to insert the dynamic trader in the process. He or she has to compute hedging ratios and value derivatives relative to other derivatives only to remain afloat in the market and to earn the right to recalibrate the tool the next day.

The Market as a Category of Thought

Of course my whole discourse presupposes the existence of the market. I won't call this a "theory" because contingent claims exist in practice, not in theory, and so does the market. As a matter of fact, I am trying to avoid theory as much as I can!

You observe that my entire "theory" presupposes the existence of the market and you fear this might be a weakness, as if the market was too special or too fragile and as if, by contrast, probability was something altogether more general and more established. You argue that the market is just a social phenomenon, as such relative and situated, whereas probability is abstract, pure, and metaphysical.

My whole point is precisely to argue to the contrary. Indeed, why wouldn't probability itself be considered a social construct? (There are even philosophers who argue that scientific theories as a whole or even reality, as a metaphysical concept, are just social constructs.) Why would probability be more general, more abstract, and somehow purer than the market? Conversely, why wouldn't the market be considered metaphysical, i.e., a pure category of thought, just like probability?

We can simply define the market as the place where contingent claims get prices attached to them. Why would such a place sound stranger and more "improbable" than the "place" where states of the world are assigned numbers that we call their "probabilities"?

Who assigns probabilities anyway, and what does probability mean as a matter of fact? Ironically, probability is in fact philosophically defined after price. Subjective probability is defined by de Finetti as the odds that a "banker" is supposed to quote for you in order to bet on the outcome of a certain event, whose probability we just define as these odds. As for objective probability, it is defined by von Mises as the limiting frequency of the occurrence of the event whose probability we wish to define within a perfectly random sequence, or as the fair price of a lottery in the long run. The infinite "random sequence" or the "long run" in question are supposed to be truly random, i.e., unbiased, and "true randomness" is then defined as a sequence of outcomes that are insensitive to gambling systems. "Banker" and "gambler": precisely the personae who deal with money and prices, not with probabilities.

The Market as a Substitute to Probability

Still, the market and the notion of price may strike us as depending too much on human beings, i.e., on two partners exchanging the contingent claim at that price, whereas probability is just independently found in nature, attaching to the thing in itself. Well, is it really? Nobody can observe probability. We only observe statistical regularities.

Statistical laws are empirical laws. However, to postulate a random generator, or a probability for the single case (what Popper calls "propensity"), is a metaphysical move, what philosophers call a reification. You can think of it as a shortcut. Instead of thinking of the whole statistical population exhibiting the given distribution, we postulate a random generator that generates each individual in turn, under a probability distribution bearing the same moments as those we have inferred statistically from the population. However, nothing grants this metaphysical extrapolation. We shouldn't forget Nassim Taleb's criticism at this juncture. There is no finite amount of statistical observations that can permit us to pin down the probability distribution that we have assumed exists behind the scenes. We really have to postulate one.

The philosophical concept of probability may have never come into existence. Ian Hacking dates its emergence back to the 17th century (and its use in gambling and dice) and its consolidation to the 19th century (in statistics). Mathematical probability theory by Kolmogorov is in fact only measure theory, bearing on set theory. It certainly has axiomatized probability calculus, but it has nothing to say about "physical" probability.

Stochastic processes are very well defined mathematically. They are the expression of probability theory at its finest. But what pure stochastic processes do we really know to exist physically? Brownian motion (of the pollen particle) is just the phenomenological summary of a multitude of invisible shocks occurring between the particle and the molecules of the liquid in which it is immersed. It is not pure. It is not a real stochastic process. As for the Brownian motion of market prices, it is no less the summary of a multitude of minute causes and transactions.

I seriously ask why the material couple formed by contingent claim and price cannot replace the metaphysical couple formed by state of the world (i.e., possibility) and probability. It is not probability that inspired us to write contingent claims in ever more complex shapes. Whoever wrote the *n*th complex contingent claim must have just had in mind the complex condition where it would pay off. He or she never thought of probability. He or she only thought of writing this payoff.

On the contrary, it seems to me that whoever thinks of a complex (abstract) state of the world and tries to figure out its probability is in fact somebody who just stops short of writing the corresponding contingent claim. My advice to him or her: just write down this complex condition; just materialize it in writing; get it out of your head! Then my whole idea is that, as he or she in effect writes it down, the material sheet on which he or she writes it is *ipso facto* meant to be exchanged in a market.

Don't ask why or how. This idea is both the definition of the market and the definition of writing. One always writes for somebody else (e.g., one writes a letter, or a testament). When you write something, you let go of it, you let it float; and this idea just means that the written contingent claim admits of a price. Contingency is written, *therefore* it is exchanged. Both the writing and the exchange are material. Possibility, by contrast, is immaterial, and so is probability.

Nonexistence of the Random Generator

The prices of contingent claims have to verify the nonarbitrage principle. A theorem states that nonarbitrage is enforced as soon as a "pricing kernel" exists. This means that the prices of contingent claims that are written on an underlying have to be expressed as the discounted mathematical expectation of their payoff under a certain probability distribution of the underlying (a.k.a. risk-neutral probability distribution). This theory doesn't mean that the underlying prices are in effect generated by this probability distribution, or that probability even exists! All that it means is that the pricing operator should be positive and linear. As a matter of fact, when probability itself is defined after price (as we saw with de Finetti), it is specified that the famous "banker" quoting those odds has to make sure he or she is not arbitraged away!

In other words, it is the market prices of the contingent claims that "generate" the prices of their underlying, not the other way around. What I mean by this is just the observation of what the market makers of contingent claims do every day when they use their pricing models. They calibrate them to the prices of vanilla options, barrier options, variance options, etc., and they infer this famous risk-neutral probability. The next day they recalibrate, possibly enlarging the pricing model to accommodate the prices of more complex contingent claims. The risk-neutral distribution they infer thus keeps shifting. This change disrupts the whole notion of a random generator, which would stably generate the prices of the underlying.

As a matter of fact, the existence of a market of contingent claims, where none should be redundant and all must independently trade, is a direct proof of the nonexistence of a random generator for the underlying. In other words, it is a direct proof of the nonexistence of states of the world. The whole metaphysical notion of possibility has to go away, together with the notion of probability.

The Step Beyond

A further objection to my "theory" is that the market price can replace probability only in the market-specific situation. How could the market replace probability when dealing with the probabilities of events other than the triggering of payoffs of contingent claims?

My objection to the objection: Is it really probability that is applied to those "foreign" events? Surely probability can be applied to dice, to roulette wheels, to the motion of the pollen particle, and to population statistics because this application is circular. Indeed, these are precisely the statistical phenomena from which probability emerged as a concept to begin with. However, the real challenge for probability is to apply to real events (happening outside the casino or the tables of actuarial science), to singular events that are in no way statistical. And what are "real events"? They are precisely events that disturb the range of possibilities on which probability was supposed to be defined—what Taleb calls "black swans."

The whole charge that Taleb is mounting against probability and its theorists is precisely concerned with such events. Badiou is the main philosopher of the event who formalized, more than 20 years ago, all that Taleb is trying to say about black swans. At no point does Badiou mention probability. Deleuze is the other, perhaps less formalistic, philosopher of the event. He speaks of the Nietzschean dice throw (which I am sure Deleuze would agree is the market of contingent claims, if only he had known it like I do) and of the "empty square" that keeps redistributing the probability distributions (and which I am sure Deleuze would agree corresponds to what I call "recalibration").

All I am saying, in the end, is that because the market price is the translation of the contingent claims without the intermediary of states of the world and their artificial delimitation—and because the whole trick of writing the formulas of contingency over the contingent claims amounts to getting rid of the underlying states—the market and price may just be the substitute of probability that is needed in such situations! True, the task remains to see how the notion of price can possibly be generalized to domains other than financial.

I think we must first try to generalize the idea of the writing of contingent claims. I believe the gist lies in writing. The written stuff people exchange when dealing with events that are more grandiose than financial payoffs are books. And the corresponding market is the sphere of thought at large. The only problem is that books are priceless.

Going Forward

In conclusion, I am really, seriously arguing for a direct passage from contingent claims (i.e., material writing instead of immaterial possibilities) to prices without ever mentioning probability. Surely probability can help us statistically analyze past prices, i.e., infer probability distributions from the observed statistical regularities. However, this reasoning is circular, as I have said, because "objective probability" is just another word for the statistical regularity that the metaphysician dreams of elevating to the status of law of nature. However, probability can be of no help with regard to future prices because random generators simply don't exist in the market. Therefore, probability can in no way help us make educated guesses.

The probabilistic tools that we use in the pricing of contingent claims are always used in reverse. That is to say, their only use is to calibrate them to the market prices of contingent claims and to thus infer the risk-neutral probability that allows us to price other stuff without creating arbitrage opportunities. In other words, the tool is just a sophisticated interpolator and extrapolator of prices of contingent claims.

Finally, a word about how this duality between price and probability, or between forward and backward, has first struck me. (Probability is the backward view, of course, and price is the forward view.) Actually, this duality is well known to quantitative analysts who deal with the so-called backward and forward pricing equations.

A backward partial differential equation (typically the Black-Scholes-Merton equation) produces the price of a single call option (say) for all times t lying between its maturity and the present time and for all spot levels S. It just computes the discounted mathematical expectation of its payoff in all those states of the world (S, t).

By contrast, a forward equation produces the prices of all call options of different maturities T and different strikes K as seen from the present spot and present time, i.e., in the present market. My contention is that the forward equation, although based on an algorithm and a mathematical formalism that are just the reciprocal image of the backward equation, is not a computation of mathematical expectations. In fact, it is not based on probability. Its state variables are not states of the world (S, t) but precisely the marks (K, T) that are written on the contingent claims, namely, their strike and maturity. The forward equation is the instant view of the instant market. You have to use a different forward equation if you wish to compute the call prices from a different spot and different time, i.e., in a different market situation.

There is nothing to stop the call prices from moving completely unpredictably between two computations. The state of the market at a given time t and for an underlying price S is completely unrelated to its state at a different time t' and for an underlying price S'. The two market states or situations are not connected by a common tree, as is the case in the backward equation. On the other hand, forward equations are better suited for calibration and recalibration to the market prices because they instantly give, in a single solve, the prices of call options of all strikes and maturity dates. They correspond better to the view that the market prices of derivatives are the actual reference, as opposed to a hypothetical random generator for the underlying, or again, that price is the reality, as opposed to probability, with the consequence that two instances of calibration can be unrelated.

References

Ayache, E., *The Blank Swan: The End of Probability*, Chichester: John Wiley & Sons Ltd, 2010.

Badiou, A., *Being and Event*, translated by Oliver Feltham, London and New York: Continuum, 2009.

de Finetti, B., *Theory of Probability,* Vols. 1 and 2, English translation, New York: Wiley, 1974.

Deleuze, G., *The Logic of Sense*, translated by Mark Lester, New York: Columbia University Press, 1990.

Hacking, I., *The Emergence of Probability*, Cambridge: Cambridge University Press, 1975.

Kolmogorov, A. N., *Foundations of the Theory of Probability,* New York: Chelsea Publishing Company, 1956.

Popper, K. R., *Realism and the Aim of Science,* Totowa, N.J.: Rowman and Littlefield, 1983.

Taleb, N. N., *The Black Swan: The Impact of the Highly Improbable,* New York: Random House, 2007.

von Mises, R., *Probability, Statistics and Truth,* 2nd revised English edition, London: Allen and Urwin, 1957.

An abc *Proof Too Tough Even for Mathematicians*

Kevin Hartnett

On August 30, 2012, a Japanese mathematician named Shinichi Mochizuki posted four papers to his faculty website at Kyoto University. Rumors had been spreading all summer that Mochizuki was onto something big, and in the abstract to the fourth paper Mochizuki explained that, indeed, his project was as grand as people had suspected. Over 512 pages of dense mathematical reasoning, he claimed to have discovered a proof of one of the most legendary unsolved problems in math.

The problem is called the *abc* conjecture, a 27-year-old proposition considered so impossible that few mathematicians even dared to take it on. Most people who might have claimed a proof of *abc* would have been dismissed as cranks. But Mochizuki was a widely respected mathematician who'd solved hard problems before. His work had to be taken seriously.

Even so, it raised an immediate problem. As a contributor named James Taylor wrote in a post to Math Overflow, a discussion board popular in the tight-knit world of higher mathematics, the question amounted to this: Could anyone explain the philosophy behind Mochizuki's work? The answer was a resounding "no."

In most fields, including math, researchers move together. They build on one another's work and cluster around solving big problems, the way physicists did in recent years with the search for the Higgs boson.

Mochizuki was different. Depending on how you calculate it, he'd been working on a proof of *abc* entirely by himself for nearly 20 years. During that time, he'd constructed his own mathematical universe and populated it with arcane terms like "inter-universal Teichmüller theory" and "alien arithmetic holomorphic structures."

Other mathematicians knew he was inventing some exotic and potentially brilliant mathematical machinery, but they had largely ignored his work, deeming it too abstruse and not worth the effort to try and understand.

Now the normally ordered world of higher mathematics is about to do something extremely unusual, plunging into a realm of abstraction and logic that even specialists don't understand. It's possible that they'll be stumbling down a colossal blind alley. It's also possible that the exploration of Mochizuki's work will change mathematics forever. If Mochizuki is right, he will have done much more than proven the *abc* conjecture: This quiet, 43-year-old native of Tokyo will have invented a whole new branch of math and transformed the way we understand numbers.

The *abc* conjecture is a young problem in mathematics, first proposed in 1985 by the mathematicians Joseph Oesterlé and David Masser to describe the relationship between three numbers: *a*, *b*, and their sum, *c*. The conjecture says that if those three numbers don't have any factors in common apart from 1, then the product of their distinct prime factors (when raised to a power slightly greater than one), is almost always going to be greater than *c*.

The conjecture intrigues mathematicians because, according to traditional thinking, there shouldn't be any connection between the prime factors of *a* and *b* and the prime factors of their sum. If the *abc* conjecture is true, it suggests that there's some hidden property of prime numbers that extends down deeper than we've been able to perceive. With his proof, Mochizuki claims to have put his finger on a previously imperceptible thread running through the ordinary

"The *abc* conjecture is somewhat mysterious," says Minhyong Kim, a professor at the University of Oxford and the Pohang University of Science and Technology and a longtime acquaintance of Mochizuki. "It is really saying that the process of adding and multiplying ordinary numbers constrains each other in a subtle but precise way. How can there be something new to say about the relation between addition and multiplication? But there seems to be."

At the time Mochizuki began working intently on *abc*, the problem was barely a decade old and little progress had been made, meaning Mochizuki was essentially cutting his own trail. The most notable work on *abc* had come from another well-regarded Japanese mathematician, Yoichi Miyaoka, who in 1988 claimed to have proven the conjecture. Miyaoka's mathematics were considered beautiful and elegant, but ultimately the proof collapsed when other mathematicians checked his work and found serious flaws. Mochizuki's work amounts to only the third serious attempt to prove the conjecture since then.

In order to understand why Mochizuki's situation is unusual, it's useful to compare it to the two biggest discoveries in math in the last 20 years: the British number theorist Sir Andrew Wiles's proof of Fermat's Last Theorem (he was knighted for his accomplishment) in 1995 and Russian mathematician Grigori Perelman's proof of the Poincaré conjecture in 2003. Hailed as singular discoveries, these proofs transformed their authors into the two most famous mathematicians in the world (alongside, perhaps, John Nash of *A Beautiful Mind* fame). But their work built from a base of well-understood mathematics, following routes that others had already speculated could lead to proofs. As a result, the mathematical community was able to verify Wiles's and Perelman's proofs in relatively short order.

Not so with Mochizuki's proof.

Mochizuki made a name for himself in his 20s based on a number of significant contributions to a relatively new, complex subfield of arithmetic geometry known as anabelian geometry. In 1998 he received one of the highest honors in math—an invitation to address the quadrennial International Congress of Mathematicians. He gave his talk in August 1998 in Berlin, then effectively went to ground, disappearing for 14 years to work on *abc*.

"I guess he wanted to work on a problem worthy of putting his full powers on," says Jeffrey Lagarias, a professor of math at the University of Michigan, "and the *abc* conjecture fit beautifully."

Before mathematicians can even start to read the proof, or understand his four papers, they need to wade through 750 pages of Mochizuki's incredibly complicated foundational work in anabelian geometry. At the moment, there are only about 50 people in the world who know anabelian geometry well enough to understand this preliminary work. Then, the proof itself is written in an entirely different branch of mathematics called "inter-universal geometry" that Mochizuki—who refers to himself as an "inter-universal geometer"—invented and of which, at least so far, he is the sole practitioner.

"Mathematics is very painful to read, even for mathematicians," says Kim, explaining why vetting Mochizuki's proof poses such a formidable task. "Most mathematicians, even people who have the necessary background knowledge in general arithmetic geometry, it's hard to convince them to put in the energy and time to read the paper."

Mochizuki is known as an uncommonly clear and poetic writer for a mathematician, but well-established mathematicians don't have much incentive to put in the years it would take to understand his work: Their research programs are set and unlikely to change dramatically in response. But a handful of up-and-coming mathematicians have seized on Mochizuki's potential proof as a chance to get in on the ground floor of a possible new field.

Jordan Ellenberg, a professor of mathematics at the University of Wisconsin, is one of them. He's spent the last two months trying to absorb Mochizuki's ideas. He's far from convinced that the proof works, but he's intrigued by its immense possibility.

"When Wiles proved Fermat," Ellenberg says, "people were energized to understand his work because they knew he could only have done that if he had understood something true and new about arithmetic. A whole field of mathematics and dozens of people's careers blossomed out of Wiles's original paper. That's the best-case scenario with Mochizuki; that's the hope."

Vesselin Dimitrov, a graduate student at Yale University, has been concentrating on reading Mochizuki's preliminary writing as preparation for reading the proof. In a series of e-mails, he explained that he's drawn by both the challenge of the *abc* conjecture and the elegance of Mochizuki's thinking. "Reading through Mochizuki's world," Dimitrov writes, "I am much impressed by the unity and structural coherence that it exhibits."

Dimitrov stressed that it's too early to predict whether Mochizuki's proof will stand up to the intense scrutiny coming its way. In October he and a collaborator, Akshay Venkatesh at Stanford University, sent a letter to Mochizuki about an error they found in the third and fourth papers of the proof. In response, Mochizuki posted a reply to his website acknowledging the error but explaining that it was minor and didn't affect his conclusions. He is expected to post a corrected version of his proof by January.

The math community has reacted to Mochizuki's proof with equal parts hope and skepticism, though few mathematicians are willing to discuss their doubts on the record out of respect for Mochizuki and a desire not to prejudge the vetting process.

Mathematicians speak of a "brick wall" in mathematical reasoning that has thwarted previous attempts to solve *abc*. "Before Mochizuki came along, this problem was viewed as utterly, hopelessly intractable and out of reach, like an out-of-the-solar-system kind of situation," says Lagarias. In that light, any supposed proof was bound to be greeted with some doubt.

Another source of skepticism is the potential expansiveness of Mochizuki's accomplishment. It has long been understood by mathematicians that any proof of *abc* would have the effect of simultaneously proving four other theorems (the work of Roth, Baker, Faltings, and Wiles) that stand among the most celebrated achievements in math in the past half-century. If Mochizuki has found a way to subsume those monumental results into a single formula, his work would take its place alongside equations like Einstein's $E = mc^2$ and the inequality behind Heisenberg's uncertainty principle in terms of its sheer explanatory power. To many, such a discovery seems too good to be possible.

Minhyong Kim thinks that the initial reaction to Mochizuki's work comes from something else. "Frankly, there are many people who express skepticism because they look at it and they can't understand what's going on, and of course when you can't understand something the most natural initial response is to be skeptical."

Meanwhile, mathematicians like Ellenberg and Dimitrov will continue to poke at Mochizuki's proof, looking for openings, raising questions, and translating Mochizuki's ideas into terms that a wider circle of mathematicians can understand.

If Mochizuki's work makes it through these informal early checks, work groups and conferences will be organized around his ideas. The Clay Mathematics Institute at Oxford has already expressed interest in sponsoring one such workshop. Further down the line—perhaps a year from now, if Mochizuki is able to build up sufficient trust with his peers—Mochizuki will submit his work for journal publication, and it will be sent out for peer review.

Mochizuki's reputation as a gifted mathematician will survive even if his proof turns out to be wrong. But the same cannot be said for his work of the past decade. Mathematicians are tantalized by the possibility of a proof of *abc*, but if an error is found early in the vetting process, the math world will likely move on without bothering to explore the rest of the mathematical universe that Mochizuki has created. Mochizuki has almost certainly made significant discoveries—but without the allure of a proof, it's possible that no one will take the time to understand them.

Mochizuki, who is known as a shy person and has declined interview requests since publishing the proof in August, will have an important role to play in all of this, answering queries and explaining his work to a math community from which he long ago parted ways.

And this situation, Kim thinks, might pose the greatest challenge of all. "When you've been wrapped up in your own research program for a long time sometimes you lose a sense of what it is that other people don't understand," he says. "Other people feel quite mystified as to what he's doing, and part of him, I suspect, doesn't quite understand why."

Contributors

Elie Ayache is CEO and co-founder of ITO 33, a Paris-based financial software company and a leading specialist in the pricing of convertible bonds, in the equity-to-credit problem, and more generally, in the calibration and recalibration of volatility surfaces. Ayache was trained as an engineer at l'École Polytechnique of Paris, from which he graduated in 1987. He then pursued a career of option market-maker on the floor of MATIF (1987–1990) and LIFFE (1990–1995). Later he turned to the philosophy of probability (DEA at la Sorbonne in 1995) and to the technology of derivative pricing (creating ITO 33 in 1999). Ayache published many articles in the *Wilmott* magazine (from 2001) on the philosophy of contingent claims, drawing both on his philosophical background and on his experience as a financial engineer and technology developer. His research culminated in *The Blank Swan: The End of Probability* (Wiley 2010), a book whose purpose is to free the pricing of derivatives, and more generally the concept of price and the market, from the probability paradigm.

Sian L. Beilock is a professor of psychology and a member of the Committee on Education at The University of Chicago. Her research program sits at the intersection of cognitive science and education. She explores the cognitive and neural substrates of skill learning, as well as the mechanisms by which performance breaks down in high-pressure situations. Dr. Beilock uses converging methodologies in her research—ranging from behavioral performance measures (e.g., reaction time and accuracy), to physiological measures of stress (e.g., salivary cortisol), to neuroimaging techniques (e.g., fMRI). In addition to answering basic questions about cognition, the goal of her research program is to inform educational practice and policy.

Jim Bennett is a visiting keeper at the Science Museum, London, having retired in 2012 from his position as director of the Museum of the History of Science at the University of Oxford, where he also held the title of Professor of the History of Science. In 2001 he was awarded the Paul Bunge Prize by the German Chemical Society for work in the history of scientific instruments.

Philip J. Davis holds a doctoral degree from Harvard and is currently professor emeritus in the Division of Applied Mathematics, Brown University. He is known for his work on numerical methods and approximation theory, as well as for his writings on the history and philosophy of mathematics. His *Mathematical Experience*, written jointly with Reuben Hersh, in 1981 won a National Book Award in Science.

Kelly Delp received her Ph.D. from the University of California, Santa Barbara. She is currently an assistant professor in the Mathematics Department at Buffalo State College and will be moving to Ithaca College in the fall of 2013. Her primary mathematical interests are in geometric topology. She enjoys making things, mathematical and otherwise. When visiting Cornell University in the spring of 2010, she began a project with Bill Thurston building surfaces out of paper and craft foam. She presented this work at the 2011 Bridges Mathematics and Art conference, where she gave the keynote address.

Prakash Gorroochurn is an assistant professor in the Department of Biostatistics at Columbia University. He is also a statistical consultant in the School of Social Work. He is an associate editor of the journal *BMC Genetics,* and his main areas of research interest include mathematical population genetics, genetic epidemiology, and the history of probability and statistics. His book *Classic Problems of Probability* won the 2012 PROSE Award for Mathematics from the American Publishers Awards for Professional and Scholarly Excellence.

Gregory Goth is a freelance journalist in Oakville, Connecticut. He has been a reporter and editor at weekly and daily newspapers in upstate New York, a staff editor for the former *LACMA Physician,* a medical policy and economics magazine published by the Los Angeles County Medical Association, and also a staff editor for *Computer,* the flagship magazine of the IEEE Computer Society. As a contributing editor, he has had work published in *IEEE Software, Computing in Science and Engineering, IEEE Internet Computing, Communications of the ACM,* and numerous other publications. He is currently a contributing editor for *Health Data Management* magazine's Healthcare Innovation Center, a website that chronicles the latest advances in healthcare networking infrastructure, clinical technologies, and where the two intersect.

Renan Gross is currently studying for his undergraduate degree in mathematics and physics at the Technion—Israel Institute of Technology. His interests include diverse topics from many fields, including computability and the origin of life. He also dabbles in science education and has mentored in an international youth science camp, Summer School of Science. Programming is one of his greatest passions, and he enjoys both writing programs to help his studies as well as learning new paradigms. In his spare time he plays the piano, sometimes with an ensemble, and enjoys cycling and slacklining.

Kevin Hartnett writes the Brainiac column for the *Boston Globe*'s Sunday *Ideas* section, where he covers news from the worlds of art, science, literature, and history. His writing as a freelance journalist has also appeared in *The Washington Post* and *The New York Observer*, among other publications. He lives in Columbia, South Carolina, with his wife and two young children.

Soren Johnson is an independent game designer. Previously he was a design director at Zynga, working on browser-based strategy games. In 2007, Johnson joined EA Maxis to work on *Spore* as a senior designer and programmer. While at EA, he also was the lead designer on *Dragon Age Legends*. Before that, he spent seven years at Firaxis, where he was the project lead and lead designer and programmer for Sid Meier's *Civilization IV*. He also programmed the artificial intelligence and was codesigner of *Civilization III*. Johnson writes a design column for *Game Developer Magazine* and is a member of the GDC advisory board. He holds a bachelor's degree in history and a master's degree in computer science from Stanford University. His thoughts on game design can be found at www.designer-notes.com.

Donald E. Knuth is a professor emeritus of computer programming at Stanford University. Among his books are four volumes (so far) of *The Art of Computer Programming*, five volumes of *Computers & Typesetting*, nine volumes of collected papers, and a nontechnical book entitled *3:16 Bible Texts Illuminated*. His software systems TeX and METAFONT are extensively used for book publishing throughout the world. He received the Turing Award from the Association for Computing Machinery in 1974, the National Medal of Science from President Carter in 1979, the Steele Prize from the American Mathematical Society in 1986, the Kyoto Prize from the Inamori Foundation in 1996, and the Frontiers of Knowledge Award from the BBVA Foundation in 2010.

Bob (David R.) Lloyd was a professor of general chemistry at Trinity College Dublin for 23 years until his retirement in 2000; previously he worked at the University of Birmingham. He is a fellow of the (London) Institute of Physics, and a member of the Royal Irish Academy. Most of his research work was concerned with the electronic structure of molecules, both free and interacting with metal surfaces, studied by various photoemission methods. In retirement, he has been working on Plato's geometrical theory of the elements, in particular on the importance of symmetry for an understanding of Plato's constructions. He has publications about this topic in journals devoted to classical studies, and to the philosophy of chemistry, and he has also found errors in more recent discussions of the Platonic solids, including the paper that appears here.

Erin A. Maloney is a postdoctoral scholar in psychology at The University of Chicago. She received her Ph.D. in cognitive psychology from the University of Waterloo in Ontario, Canada, in 2011. Her research program sits at the intersection of cognitive psychology, social psychology, and education. Maloney explores the cognitive and social factors that influence performance in mathematics, focusing on the role of numerical competency, anxiety, and stereotypes in math achievement. Her research takes place both in the lab studying adults and in the classroom studying young children, their teachers, and their parents. In addition to answering basic questions about cognition, with her research, Maloney aims to inform educational practice and policy.

John Pavlus is a filmmaker and writer interested in science, math, design, technology, and other ways that people make things make sense. His work has appeared in *Scientific American, Wired, Fast Company, Technology Review,* and elsewhere. He creates original short films and documentaries with partners including NPR, Autodesk, The Howard Hughes Medical Institute, and *The New York Times Magazine* via his production company, Small Mammal. He lives in Portland, Oregon.

Sir Roger Penrose is Emeritus Rouse Ball Professor of Mathematics at Oxford University. He received the 1988 Wolf Prize in physics for his contribution with Stephen Hawking in understanding the universe. He is the author of several books, including *The Nature of Space and Time* (Princeton University Press).

Frank Quinn has contributed extensively to the study of manifolds, stratified sets, and algebraic K-theory. He is best known for his work on 4-manifolds and is a leading practitioner of visualizing subtle things in four dimensions. His unorthodox views on mathematics might be influenced by this. He has been a professor of mathematics at Virginia Tech for more than 35 years.

Fiona Donaghey Ross teaches fine art at the University of Richmond. Her paintings, drawings, and sculptures are in the collections of Capital One, Wachovia Securities, Markel Corporation, and the Republic of Ireland. Her works have been featured in many national and international exhibitions, including Zone: Chelsea in New York City, the Seoul Hae-Tae Gallery in South Korea, and the Shang Shang Gallery in Beijing, China. She is a recipient of the

Virginia Museum of Fine Arts Professional Artist Grant and is a fellow at the Virginia Center for Creative Arts. Ross's work has been published in books, such as *Ceramics: Mastering the Craft*, and magazines, including *Korean Ceramics Monthly* and *New American Paintings*. She is represented in Richmond, Virginia, by the Page Bond Gallery.

William T. Ross is the Roger Francis and Mary Saunders Richardson Chair in Mathematics, and chair of the Mathematics and Computer Science Department at the University of Richmond. He is the author of more than 30 research papers, as well as four monographs on complex analysis and operator theory. He is also a painter.

Charles Seife, a professor of journalism at NYU's Arthur L. Carter Journalism Institute, has been writing about physics and mathematics for two decades. He is the author of five books, including *Zero: The Biography of a Dangerous Idea* (2000), which won the 2000 PEN/Martha Albrand Award for First Nonfiction; *Sun in a Bottle: The Strange History of Fusion and the Science of Wishful Thinking* (2008), which won the 2009 Davis Prize from the History of Science Society; and *Proofiness: The Dark Arts of Mathematical Deception* (2010).

Anna Sfard is a professor of mathematics education at the University of Haifa. She served as Lappan-Philips-Fitzgerald Professor at Michigan State University and is a visiting professor in the Institute of Education, University of London, in the United Kingdom. Her research, focusing on the development and role of mathematical discourses in individual lives and in the course of history, has been summarized in her book, *Thinking as Communicating*. She is the recipient of the 2007 Freudenthal Award.

Daniel S. Silver is a professor of mathematics at the University of South Alabama. Much of his published research explores the relationship between knots and dynamical systems. Other active interests include the history of science and the psychology of invention. He has contributed articles of general interest to *American Scientist* and *Notices of the American Mathematical Society*.

Ian Stewart is an emeritus professor of mathematics at the University of Warwick. His research areas include pattern formation, networks, and biomathematics. He is a fellow of the Royal Society and was awarded its Michael Faraday Medal for the public understanding of science. He has written many popular mathematics books, including *17 Equations that Changed the World* and *Professor Stewart's Cabinet of Mathematical Curiosities*.

Terence Tao was born in Adelaide, Australia, in 1975. He has been a professor of mathematics at UCLA since 1999, after completing his Ph,D, under Elias Stein at Princeton in 1996. Tao's areas of research include harmonic analysis, partial differential equations, combinatorics, and number theory. He has received a number of awards, including the Salem Prize in 2000, the Bochner Prize in 2002, the Fields Medal and the SASTRA Ramanujan Prize in 2006, the MacArthur Fellowship and Ostrowski Prize in 2007, the Waterman Award in 2008, the Nemmers Prize in 2010, and the Crafoord Prize in 2012. Terence Tao also currently holds the James and Carol Collins Chair in Mathematics at UCLA, and is a fellow of the Royal Society, the Australian Academy of Sciences (corresponding member), the National Academy of Sciences (foreign member), and the American Academy of Arts and Sciences.

Notable Texts

The most difficult part of reaching the final content for each volume of this series is to leave out worthy pieces that ultimately cannot be included. I list below entries I considered for selection this year but that did not make it into the book because of various constraints, mostly related to the space available and/or to issues of copyright.

Notable Articles

Alexanderson, Gerald L. "John Wallis and Oxford." *Bulletin of the American Mathematical Society* 49.3(2012): 443–446.

Alms, Jeremy F., and David A. Andrews. "Solutions in Search of Problems." *Notices of the American Mathematical Society* 59.7(2012): 963–964.

Alperin, Roger C., Barry Hayes, and Robert J. Lang. "Folding the Hyperbolic Crane." *Mathematical Intelligencer* 34.2(2012): 38–49.

Aron, Jacob. "A Mathematical Universe Is Born." *New Scientist* 215(Sep. 15, 2012): 6.

Barrett, Linda K., and B. Vena Long. "The Moore Method and the Constructivist Theory of Learning: Was R. L. Moore a Constructivist?" *PRIMUS* 22.1(2012): 75–84.

Beer, Gillian. "Alice in Time." *Nature* 479(Nov. 3, 2012): 38–39.

Bishop, Alan J. "From Culture to Well-Being." *Zentralblatt für Didaktik der Mathematik* 44(2012): 3–8.

Borjas, George J., and Kirk B. Doran. "The Collapse of the Soviet Union and the Productivity of the American Mathematicians." *Quarterly Journal of Economics* 127.3(2012): 1143–1203.

Bueno, Otávio, and Steven French. "Can Mathematics Explain Physical Phenomena?" *British Journal of the Philosophy of Science* 63(2012): 85–113.

Burgiel, Heidi, and Matthew Salomone. "Logarithmic Spirals and Projective Geometry in M. C. Escher's *Path of Life III*." *Journal of Humanistic Mathematics* 2.2(2012): 22–35.

Clements, Douglas H., and Julie Sarama. "Mathematics Learning, Assessment, and Curriculum." In *Handbook of Early Childhood Education*, edited by Robert C. Pianta, New York: Guildford Press, 2012, pp. 217–239.

Conway, John, and Alex Ryba. "The Pascal Mysticum Demystified." *Mathematical Intelligencer* 34.3(2012): 4–8.

Cooper, Barry S. "Incomputability after Alan Turing." *Notices of the American Mathematical Society* 59.6(2012): 776–784.

de Freitas, Elizabeth. "The Diagram as Story." *For the Learning of Mathematics* 32.2(2012): 27–33.

Dedò, Maria, and Laura Sferch. "Right or Wrong? That Is the Question." *Notices of the American Mathematical Society* 59.7(2012): 924–932.

Denning, Peter, and Tim Bell. "The Information Paradox." *American Scientist* 100.6(2012): 470–477.

Dunlop, Katherine. "Kant and Strawson on the Content of Geometrical Concepts." *Noûs* 46.1(2012): 86–126.

Gelman, Andrew, and Eric Loken. "When We (Statisticians) Teach, We Don't Practice What We Preach." *Chance* 25.2(2012): 24–25.

Glaz, Sarah. "Poetry Inspired by Mathematics: A Brief Journey through History." *Journal of Mathematics and the Arts* 5.4(2011): 171–183.

Grabiner, Judith. "Why Proof? A Historian's Perspective." In *Proof and Proving in Mathematics Education*, edited by Gila Hanna and Michael de Villiers. Dordrecht, Germany: Springer Science+Business Media, 2012.

Gray, Jeremy. "Poincaré and the Idea of a Group." *Nieuw Archief voor Wiskunde* 13.3(2012): 178–186.

Gray, Jeremy. "Poincaré Replies to Hilbert: On the Future of Mathematics ca. 1908." *Mathematical Intelligencer* 34.3(2012): 15–29.

Grünbaum, Branko. "Is Napoleon's Theorem *Really* Napoleon's Theorem?" *The American Mathematical Monthly* 119.6(2012): 495–501.

Hacker, Andrew. "Is Algebra Necessary?" *The New York Times* July 28, 2012.

Hacking, Ian. "The Lure of Pythagoras." *Iyyun—The Jerusalem Philosophical Quarterly* 61.2(2012): 103–128.

Hahn, Alexander J. "Geometric Architecture: Up and Add 'Em." *berfrois* Aug. 31, 2012.

Halpern, Diane F. et al. "Sex, Math, and Scientific Achievement: Why Do Men Dominate the Fields of Science, Engineering, and Mathematics?" *Scientific American Mind* 27(2012): 26–33.

Hayes, Brian. "Alice and Bob in Cipherspace." *American Scientist* 100.5(2012): 362–367.

Hill, Theodore P., and Erika Rogers. "Gender Gaps in Science: The Creativity Factor." *Mathematical Intelligencer* 34.2(2012): 19–26.

Johnston, Stephen. "John Dee on Geometry: Texts, Teaching and the Euclidean Tradition." *Studies in History and Philosophy of Science* 43.2(2012): 470–479.

Kehle, Paul. "Quasicrystals: Mathematicians Were There First." *Consortium* 102(Spring 2012): 13–17.

Kilpatrick, Jeremy. "The New Math as an International Phenomenon." *Zentralblatt für Didaktik der Mathematik* 44.4(2012): 563–571.

Krasner, Daniel. "A Different Path (from Mathematics to Startups." *Notices of the American Mathematical Society* 58.11(2011): 1588–1591.

Lai, Yvonne, Keith Weber, and Pablo Mejía-Ramos. "Mathematicians' Perspectives on Features of a Good Pedagogical Proof." *Cognition and Instruction* 30.2(2012): 146–169.

Lakshtanov, Evgeny, and Vera Roschina. "On Fitness in the Card Game of War." *The American Mathematical Monthly* 119.4(2012): 318–323.

Lazar, Nicole. "Big Data Hits the Big Time." *Chance* 25.3(2012): 47–49.

Lewis, Michael A. "Mathematics and *The Hunger Games*." *Journal of Humanistic Mathematics* 2.2(2012): 129–139.

Macbeth, Danielle. "Seeing How It Goes: Paper-and-Pencil Reasoning in Mathematical Practice." *Philosophia Mathematica* 20.1(2012): 58–85.

Mackenzie, Dana. "A Flapping of Wings." *Science* 335 (March 23, 2012): 1430–1433.

Mauldin, Tim. "Time and the Geometry of the Universe." In *The Future of the Philosophy of Time,* edited by Adrian Bardon. New York: Routledge, 2012.

Merow, Katharine. "Math Is My *Femme Fatale*." *Mathematical Intelligencer* 34.1(2012): 42–43.

Morgan, Candia, and Jehad Alshwaikh. "Communicating Experience of 3D Space: Mathematical and Everyday Discourse." *Mathematical Thinking and Learning* 14.3(2012): 99–125.

Muntersbjorn, Madeline. "On the Intellectual Heritage of Henri Poincaré." *British Society for the History of Mathematics Bulletin* 27.2(2012): 107–118.

Nicholson, Jason Scott. "A Perspective on Wigner's 'Unreasonable Effectiveness of Mathematics.'" *Notices of the American Mathematical Society* 59.1(2012): 38–42.

Papademetri-Kachrimani, Chrystalla. "Revisiting van Hiele." *For the Learning of Mathematics* 32.3(2012): 2–7.

Pedersen, Jean, and Tibor Tarnai. "Mysterious Movable Models." *Mathematical Intelligencer* 34.3(2012): 62–66.

Plofker, Kim. "Mathematics and Its Worldwide History." *Nieuw Archief voor Wiskunde* 13.1(2012): 18–24.

Schoenfeld, Alan H. "Problematizing the Didactic Triangle." *Zentralblatt für Didaktik der Mathematik* 44.5(2012): 587–599.

Sequin, Carlo H. "Topological Tori as Abstract Art." *Journal of Mathematics and the Arts* 6.4(2012): 191–209.

Sinclair, Nathalie, Anne Watson, Rina Zazkis, and John Mason. "The Structuring of Personal Example Spaces." *Journal of Mathematical Behavior* 30(2011): 291–303.

Sophian, Catherine. "Mathematics for Early Childhood Education." In *Handbook of Research on the Education of Young Children*, 3rd edition, edited by Olivia N. Saracho and Bernard Spodek. London, UK: Routledge, 2012, pp. 169–178.

Stillwell, John. "Poincaré and the Early History of 3-Manifolds." *Bulletin of the American Mathematical Society* 49.4(2012): 555–577.

Strayer, Daniel, and Elizabeth Brown. "Teaching with High-Cognitive-Demand Mathematical Tasks." *Notices of the American Mathematical Society* 59.1(2012): 55–57.

Strogatz, Steven. "Singular Sensations." *The New York Times* Sept. 10, 2012. This article opened a series of six remarkable weekly pieces.

Stumpf, Michael P. H., and Mason A. Porter. "Critical Truths about Power Laws." *Science* 335(Feb. 10, 2012): 665–666.

Thomson-Jones, Martin. "Modeling without Mathematics." *Philosophy of Science* 79.5(2012): 761–772.

Timmermans, Benoît. "Prehistory of the Concept of Mathematical Structure." *Mathematical Intelligencer* 34.3(2012): 41–54.

Usiskin, Zalman. "Misidentifying Factors Underlying Singapore's High Test Scores." *Mathematics Teacher* 105.9(2012): 666–670.

van Dalen, Dirk. "Poincaré and Brower on Intuition and Logic." *Nieuw Archief voor Wiskunde* 13.3(2012): 191–195.

Webb, Richard. "From Zero to Hero." *The New Scientist* 2839(Nov. 19, 2011): 41–43.

Widom, Theodore Reed. "Methodological Reflections on Typologies for Numerical Notations." *Science in Context* 25.2(2012): 155–195.

Wilkie, James E. B., and Galen V. Bodenhausen. "Are Numbers Gendered?" *Journal of Experimental Psychology: General* 141.2(2012): 206–210.

Williams, Wendy, and Stephen Ceci. "When Scientists Choose Motherhood." *American Scientist* 100(2012): 138–145.

Yong, Darryl. "Adventures in Teaching: A Professor Goes to High School to Learn about Teaching Math." *Notices of the American Mathematical Society* 59.10(2012):1408–1415.

Yoshinobu, Stan, and Matthew G. Jones. "The Coverage Issue." *PRIMUS* 22.4(2012): 303–316.

Young, Christina B., Sarah S. Wu, and Vinod Menod. "The Neurodevelopmental Basis of Math Anxiety." *Psychological Science* 23.5(2012): 492–501.

Notable Journal Issues

The following special issues of 2012 journals are likely to appeal to the reader interested in nontechnical writings and topics concerning mathematics. This list is far from being comprehensive.

"Science in the 21st Century." *Daedalus* 141.3(2012).

"John Dewey and the Child." *Education and Culture* 28.2(2012).

"The Inverse Principle: Psychological, Mathematical, and Educational Considerations." *Educational Studies in Mathematics* 79.3(2012).

"Activity Theoretical Approaches to Mathematics Classroom Practices with the Use of Technology." *International Journal for Technology in Mathematics Education* 19.4(2012) & 20.1(2013).

"Mathematical Knowledge and Its Applications." *Iyyun: The Jerusalem Philosophical* Quarterly 61.2(2012).

"Teacher Knowledge, Curriculum Materials, and Quality of Instruction." *Journal of Curriculum Studies* 44.4(2012).

"Modalities of Body Engagement in Mathematical Activity and Learning." *Journal of the Learning Sciences* 21.2(2012).

"Foregrounding Equity in Mathematics Teacher Education." *Journal of Mathematics Teacher Education* 15.1(2012).

"Improving Self-Monitoring and Self-Regulation of Learning: From Cognitive Psychology to the Classroom." *Learning and Instruction* 22.4(2012).

"Big Data." *Significance* 49.4(2012).

"Diagrams in Mathematics: History and Philosophy." *Synthese* 186.1(2012).

"New Perspectives on the Didactic Triangle: Teacher–Student–Content." *Zentralblatt für Didaktik der Mathematik* 44.5(2012).

Acknowledgments

This series of books owes its merits primarily to the authors whose contributions are included and to the original publishers; I thank them first, for cooperating in the republication in this form.

For the fourth consecutive year, Gerald Alexanderson and Fernando Gouvêa read a wider collection of articles than we present here; their opinionated, witty, learned, and at times sharply opposite comments on the texts I advanced helped me select the list I chose for the book. Yet the entire responsibility for the shortcomings you might find is mine. In the advanced stages of deciding on the final content, I also benefited, once again, from consulting with Vickie Kearn of the Princeton University Press. Quinn Fusting resolved the copyright matters; Paula Bérard copyedited the manuscript, and Nathan Carr prepared it for production. Thank you to all.

Thanks to Paul Zorn for expressing a very public appreciation of this series.

The excellent staff and services at Cornell University Library are indispensable to my work. Thank you.

After a disastrous stab at graduate studies at another university (a ruinous experience in indoctrination that overturned my optimistic preconceptions about academic inquiry and introduced me to the strangely alluring whiff of dogmatism that percolates in academic circles oblivious to evidence), after a stint as full-time stay-at-home dad (a thrilling experience that was turned against my child and me with ferocity in a family court and its acolyte offices), and after surviving unusual circumstances, I was fortunate to benefit from an excellent medium for thought and research at Cornell University. This stability not only allowed me to teach a diverse assortment of classes and to read widely across disciplines but also to be close to my daughter. I owe benevolence and understanding to many people, starting with my committee members: David W. Henderson, Anil Nerode, John W. Sipple, and Steven Strogatz whose patience I tested dearly. Once again this year, Maria Terrell assigned me an adequate teaching load, for which I am grateful. In addition, thanks to Dan Barbasch, Mary Ann Huntley, Severin Drix, Michelle Klinger, Heather Peterson, William Gilligan, and Catherine Penner for offering me to teach again for the Cornell Mathematics Outreach Programs, Cornell Summer School, and Cornell Adult University; I enjoy all of these opportunities, hopefully to the benefit of my students.

The writing seminar on mathematical topics, which I taught over the past several fall semesters, played an important role in clarifying my goals for editing this series. I designed a course unique in content, dynamic, and expectations—and I feel rewarded by the discussions and by the student reactions occasioned by it. For the chance to teach the seminar again, thanks to Katherine Gottschalk, Paul Sawyer, David Faulkner, and Bruce Roebal of Cornell University's Knight Institute for Writing in the Disciplines.

This brings me to express my gratitude to the late William Thurston. I am privileged to have been among the people touched by his generosity. He was the first guest to a class meeting in my seminar and kindly agreed to come once every semester, even after he fell gravely ill. Each time, especially the last time, he conveyed the overwhelming impression that mathematical thinking can never be laid out in full; that behind the conventions captured by precise definitions, and beyond the apparent clarity of mathematical reasoning, there are endless layers still to explore; and that a mathematician thrives in the community of like-minded people, only to become different from all of them. Three years ago, despite his numerous commitments, Professor Thurston graciously accepted to write the foreword for the opening volume in this series. Now that he is no longer with us, I dedicate this volume to his memory.

Thanks to Fangfang for caring, from near and from afar.

Finally, yet most importantly, thanks to my daughter Ioana for her blissful cheerfulness. Indoors and outdoors, at home, in libraries, and during our travels, she has often seen me shuffling and reading the articles, books, and notes that led to this volume.

Credits

"The End of Probability" by Elie Ayache. Previously published in *Wilmott* (November 2010): 40–44. Copyright © 2010 John Wiley & Sons Ltd. Reprinted with permission of John Wiley & Sons Ltd.

"Early Modern Mathematical Instruments" by Jim Bennett. Previously published in *Isis* 102.4 (2012): 697–705, University of Chicago Press. © 2011 by The History of Science Society. All rights reserved. Reprinted by permission of the University of Chicago Press.

"The Prospects for Mathematics in a Multi-Media Civilization" by Philip J. Davis. Originally published in *Mathamtics Everywhere*, edited by Martin Aigner and Ehrhard Behrends, Providence, RI: American Mathematical Society, 2010. © 2010. Reprinted by kind permission of the American Mathematical Society and the author.

"High Fashion Meets Higher Mathematics" by Kelly Delp. Previously published in *Math Horizons* 20.2(2012): 5–10. Reprinted by kind permission of the Mathematical Association of America and the author.

"Errors of Probability in Historical Context" by Prakash Gorroochurn. Previously published in *The American Statistician*, Volume 65, Issue 4, (2011) pp. 246–254. Copyright 2011 from "Errors of Probability in Historical Context" by Prakash Gorroochurn. Reproduced by permission of American Statistical Association, http://www.amstat.org.

"Degrees of Separation" by Gregory Goth. Previously published in *Communications of the ACM* 55.7(2012): 13–15. Copyright © 2012 Association for Computing Machinery, Inc. Reprinted by permission. http://doi.acm.org/10.1145/2209249.2209255

"Bridges, String Art, and Bézier Curves" by Renan Gross. This article first appeared in *Plus Magazine*, http://plus.maths.org (March 5th, 2012). Reprinted by kind permission of the author.

"An ABC Proof Too Tough Even for Mathematicians" by Kevin Hartnett. Previously published in *The Boston Globe*, November 4, 2012. Reprinted by kind permission of the author.

"Playing the Odds" by Soren Johnson. Previously published in *Games, Learning, and Society*, edited by Constance Steinkuehler, Kurt Squire, and Sasha Barab. New York: Cambridge University Press, 2012. Originally published in *Game Developer Magazine*, October 2009. Copyright © 2012 Cambridge University Press. Reprinted with permission of Cambridge University Press.

"Randomness in Music" by Donald E. Knuth. Previously published in *Selected Papers on Fun and Games* by Donald E. Knuth, Stanford, CA: CSLI Publications, 2011. Reprinted by kind permission of the Donald W. Knuth and the Center for the Study of Language and Information, Stanford University.

"How Old Are the Platonic Solids?" by David R. Lloyd. Previously published in *BSHM Bulletin: British Society for the History of Mathematics Bulletin* 27.3(2012): 131–140. Copyright © British Society for the History of Mathematics, reprinted by permission of Taylor & Francis Ltd., www.tandfonline.com on behalf of The British Society for the History of Mathematics.

"Math Anxiety: Who Has It, Why It Develops, and How to Guard Against It" by Erin A. Maloney and Sian L. Beilock. Reprinted from *Trends in Cognitive Sciences 16*(8) (2012) 404–406. Copyright © 2012 with permission from Elsevier.

"Machines of the Infinite" by John Pavlus. Previously published in *Scientific American* 307.3 (2012): 66–71. Reprinted by kind permission of the author.

"A Revolution in Mathematics? What Really Happened a Century Ago and Why It Matters Today" by Frank Quinn. Previously published in *Notices of the American Mathematical Society* 59.1(2012): 31–37. Reprinted by kind permission of the author.

"The Jordan Curve Theorem Is Non-trivial" by Fiona Ross and William T. Ross. Previously published in *Journal of Mathematics and the Arts* 5.4(2011): 213–219, Taylor and Francis. Reprinted by permission of Taylor & Francis Ltd., www.tandfonline.com.

"Randomness" by Charles Seife. Previously published in *This Will Make You Smarter*, edited by John Brockman. New York: Harper, 2012, pp. 105–108. © 2012 Random House, Inc. reprinted with permission of Random House, Inc.

"Why Mathematics? What Mathematics?" by Anna Sfard. Originally published in *The Mathematics Educator* 22.1(2012): 3–16. Reprinted by kind permission of the author and *The Mathematics Educator*.

"Slicing a Cone for Art and Science" by Daniel S. Silver. Previously published in *American Scientist* 100.5(2012): 408–415. Reprinted by permission of *American Scientist*.

"Fearful Symmetry" by Ian Stewart. Previously published in *Bletchley Park Times* (Autumn 2010): 10–12. Reprinted by kind permission of the author.

"*E pluribus unum*: From Complexity, Universality" by Terence Tao. Previously published in *Daedalus*, 141.3 (2012): 23–34. © 2012 by the American Academy of Arts and Sciences. Reprinted by kind permission of the author, *Daedalus*, and MIT Press.